Computing with Memory for Energy-Efficient Robust Systems

Somnath Paul • Swarup Bhunia

Computing with Memory for Energy-Efficient Robust Systems

Somnath Paul
Intel Labs
Hillsboro, OR, USA

Swarup Bhunia
Department of EECS
Case Western Reserve University
Cleveland, OH, USA

ISBN 978-1-4614-7797-6   ISBN 978-1-4614-7798-3 (eBook)
DOI 10.1007/978-1-4614-7798-3
Springer New York Heidelberg Dordrecht London

Library of Congress Control Number: 2014946218

© Springer Science+Business Media New York 2014
This work is subject to copyright. All rights are reserved by the Publisher, whether the whole or part of the material is concerned, specifically the rights of translation, reprinting, reuse of illustrations, recitation, broadcasting, reproduction on microfilms or in any other physical way, and transmission or information storage and retrieval, electronic adaptation, computer software, or by similar or dissimilar methodology now known or hereafter developed. Exempted from this legal reservation are brief excerpts in connection with reviews or scholarly analysis or material supplied specifically for the purpose of being entered and executed on a computer system, for exclusive use by the purchaser of the work. Duplication of this publication or parts thereof is permitted only under the provisions of the Copyright Law of the Publisher's location, in its current version, and permission for use must always be obtained from Springer. Permissions for use may be obtained through RightsLink at the Copyright Clearance Center. Violations are liable to prosecution under the respective Copyright Law.
The use of general descriptive names, registered names, trademarks, service marks, etc. in this publication does not imply, even in the absence of a specific statement, that such names are exempt from the relevant protective laws and regulations and therefore free for general use.
While the advice and information in this book are believed to be true and accurate at the date of publication, neither the authors nor the editors nor the publisher can accept any legal responsibility for any errors or omissions that may be made. The publisher makes no warranty, express or implied, with respect to the material contained herein.

Printed on acid-free paper

Springer is part of Springer Science+Business Media (www.springer.com)

Preface

Energy-efficiency and reliability have emerged as major barriers to performance scalability in nanometer regime. The situation is particularly worse in the case of data-intensive workloads that require handling large volume of data from the domain of multimedia, informatics, data mining, and security applications. In the post-Dennard scaling era, with technology scaling only providing linear reduction in energy compared to cubic reduction in previous technology nodes, there is a renewed thrust to investigate novel computing frameworks which can continue the roadmap for energy-efficient computing. Conventional computing solutions, namely, Von-Neumann machines like general-purpose processors are incapable of satisfying the energy-efficiency requirements for data-intensive applications. The attention has therefore shifted to non-Von-Neumann architectures such as graphics processing units (GPUs), field-programmable gate arrays (FPGAs), and other fixed-function and reconfigurable accelerators. A close investigation of the system-level performance and energy bottlenecks for data-intensive applications, however, reveals that managing off-chip data movement and memory access is as important as improving the efficiency of computing solutions on-chip. For these data-heavy applications, energy required in on-chip computation constitutes only a small fraction of the total energy consumption. It is primarily contributed by transportation of the data from off-chip memory to on-chip computing elements—a limitation popularly referred to as the *Von-Neumann bottleneck*. In such a scenario, improving the compute energy through parallel processing or on-chip hardware acceleration brings minor improvements to the total energy required by the system. Technology trends suggest that while on-chip integration density increases exponentially, off-chip bandwidth improves only at a linear rate over the same period of time. Hence, there is a critical need to develop novel hardware architecture that enables energy-efficient computing for diverse data-intensive applications.

To address this need, in this book, we propose a novel distributed and scalable memory-centric reconfigurable hardware architecture with associated software framework for application mapping. The computing model that is realized by the hardware architecture is referred to as *memory-based computing* or MBC. Hardware reconfigurable systems have been demonstrated to be more energy-efficient compared to software programmable systems. However, they often incur large area, delay, and power overhead due to the presence of a programmable interconnect. The memory-centric reconfigurable computing model proposed in this work mitigates this overhead by optimally mapping the input application in a spatiotemporal fashion to the underlying computing elements of this framework. The framework therefore retains the flexibility and energy-efficiency of reconfigurable hardware, while making them more scalable at advanced technology nodes. The computing elements for the proposed hardware are generic enough to collectively map applications of arbitrary size and granularity. The book provides comprehensive description on the hardware and software architecture for *memory-based computing* and presents effective circuit, architecture, and software co-optimizations, which can collectively improve the energy-efficiency of the proposed memory-centric reconfigurable architecture. We also describe the functional verifications of the applications mapped to the MBC framework using both simulation and hardware emulation-based approaches.

The proposed hardware solution can be used as a stand-alone reconfigurable framework or utilized as a co-processor for data-intensive applications. This co-processor, referred to as Memory Array centric Hardware Accelerator (MAHA), offers the following distinctive advantages compared to alternative reconfigurable accelerator platforms, e.g., FPGA, Coarse-Grained Reconfigurable Architecture (CGRA), and GPU-based acceleration: (1) it exploits high-density and low access-time/energy of nanoscale memory for storing both data and lookup table-based function responses; (2) the block-based memory architecture is used to create large number of parallel memory-centric configurable computing resources (memory-centric nano-processors, otherwise referred to as memory logic blocks or MLBs), which can be effective for massive data-parallel applications; (3) each MLB implements a distinct instruction-set architecture optimized for data-intensive applications including support for lookup operations of varying bitwidth, support for complex fused operations, and support for mixed-granular operations; and (4) an optimal hierarchical bus-based architecture that leverages on the nature of data movements across MLBs at different levels of hierarchy. Information about data movement is localized into MLBs by including it inside the instructions, which enables effective error protection. The project codevelops a software framework for application mapping into an array of MLBs from the control and data flow description of an application kernel.

The memory inside each processing element can be malleably used to either perform computation (in the form of a LUT) or store data. The framework therefore benefits a wide range of compute and data-intensive workloads. Our investigations, included in this work, demonstrate that MBC achieves significant improvement in energy-efficiency over FPGA at the same technology node. The improvement is more pronounced at scaled nodes and enhances with low-power design techniques such as voltage scaling. We establish that drastic improvements in area and energy-efficiency can be achieved when the input application is expressed as a PLA and the LUT-based implementation is replaced with a content-addressable memory (CAM)-based architecture. In the nanometer regime, memories are prone to parametric and runtime failures. In this work, we demonstrate that the effect of these parametric failures can be mitigated through intelligent resource allocation step in the MBC software flow. Runtime failures were shown to be tolerated using a novel reliability aware reconfigurable error-correction code (ECC) assignment approach. In this work, we have therefore critically addressed the challenges of improving the energy-efficiency and reliability of the MBC framework itself.

The benefits of memory-based computing becomes even more apparent for computing with off-chip nonvolatile memory arrays. This holds for conventional NAND Flash memory architecture as well as emerging nonvolatile memory technologies such as Spin Torque Transfer Memory (STTM), which are amenable to dense nonvolatile memory design. For data-intensive applications, this architecture exploits high-density nanoscale memory to process the data in situ and enables massively data-parallel processing for multimedia, signal processing, security, and informatics applications with ultra-low power dissipation. We further improve the energy-efficiency for "in-memory computing" by exploring the large design space for MAHA for a variety of data-intensive applications. This involves optimizing the nonvolatile memory organization (both NAND Flash and STTM), array of nano-processors and the interconnect hierarchy. The hardware architecture for each of the nano-processors has been optimized considering the commonly used operations for these data-intensive applications. For example, the support for fused operations as well as lookup operations of varying bitwidth enables efficient application mapping. Each MLB supports mixed-granular datapath operations including random logic functions (e.g., bit manipulation functions), which can help efficient implementation of many functions in informatics and multimedia domain. Furthermore, the routing requirements are embedded into the instructions, thus making reconfiguration process more efficient. Unlike the distributed and fine-grained configuration of FPGA, it enables us to efficiently protect the configuration bits using error-correction code (ECC). As we show in our work, migrating the data-intensive workload from on-chip compute solutions to an off-chip NVM-based computing framework can significantly improve system-level energy-efficiency. The improvement is, however, not constant across the board, but varies with the nature of the application. We therefore propose a formulation based on simple application behavior to determine a priori if an application can benefit from off-chip in-memory computing. Compared to general-purpose processors, data-intensive workloads therefore enjoy 1-1000X improvement when migrated to the off-chip MAHA framework. Existing accelerators, including conventional FPGA, CGRA, and GPU-based accelerators can deliver the desired performance benefit, but suffer from the energy barrier owing to off-chip memory access. The proposed NVM-based MAHA architecture can alleviate this bottleneck and deliver low-power system operation for data-intensive workloads.

Memory-based computing blends the benefits of spatial and temporal computing and also the goodness of software and hardware reconfigurable frameworks in a single fabric. The fact that it is memory-centric makes it particularly useful for emerging data-heavy applications like informatics and amenable to implementation using emerging dense nonvolatile memories. We also believe the new research direction put forward in this work will excite the target readership consisting of students, researchers, and practitioners and they will benefit from it. We believe that the content will also remain highly relevant in future—since current technology trends and emerging technology solutions favor the case of in-memory computing frameworks. We are very grateful to all students of Nanoscale Research Lab at Case Western Reserve University for their help and support in the course of the MBC project. We also acknowledge the contribution of Prof. Saibal Mukhopadhyay and his students from the GREEN Lab at the School of Electrical and Computer Engineering, Georgia Institute of Technology. We also remain sincerely thankful to Springer, USA and all their publication staff.

Hillsboro, OR, USA Somnath Paul
Cleveland, OH, USA Swarup Bhunia

Contents

Part I Introduction

1 Challenges in Computing for Nanoscale Technologies 3
    1.1   End of Dennard Scaling 3
    1.2   Sustaining Reliability of Operation 5
    1.3   Addressing Energy-Efficiency for Data-Intensive Workloads 6
    References 8

2 A Survey of Computing Architectures 11
    2.1   Von-Neumann and Non-Von-Neumann Architectures 11
    2.2   Reconfigurable Computing: State of the Art 12
    2.3   Challenges in Reconfigurable Hardware Design 21
    References 25

3 Motivation for a Memory-Based Computing Hardware 29
    3.1   Motivation for a New Computing Model 29
    3.2   Proposed Solution and Benefits 32
    References 33

Part II Memory Based Computing: Overview

4 Key Features of Memory-Based Computing 37
    4.1   Computing in Memory Versus Computing with Memory 37
    4.2   Highlights of Memory-Based Computing Framework 39
    References 40

5 Overview of Hardware and Software Architectures 41
    5.1   Overview of Hardware Organization and Software Flow 41
    5.2   Distinction with Existing Hardware Frameworks 42
    References 46

6 Application of Memory-Based Computing 47
6.1 Motifs 48
6.2 Contexts 48
6.3 Domains 49
References 50

Part III Hardware Framework

7 A Memory Based Generic Reconfigurable Framework 53
7.1 A Generic Reconfigurable Framework 53
7.2 A Generic Reconfigurable Framework with MBC 55
References 63

8 MAHA Hardware Architecture 65
8.1 A Typical Memory Organization 65
8.2 Instrumenting Memory for Hardware Acceleration 67
References 76

Part IV Software Framework

9 Application Analysis 79
9.1 Application Description Using a CDFG 79
9.2 Decomposition 81
9.3 Fusion 85
References 89

10 Application Mapping to MBC Hardware 91
10.1 Resource-Aware Scheduling 91
10.2 Packing of Partitions to Multi-MLB Framework 93
10.3 Placement 96
10.4 Report Generation 99
10.5 Bitfile Generation and Functional Validation 102
10.6 Design Space Exploration 103
References 104

Part V MBC Design Space Exploration

11 Design Space Exploration for MBC Based Generic Reconfigurable Computing Framework 107
11.1 Benchmarks and Experimental Setup 107
11.2 MLB Architecture Exploration 109
11.3 Comparison with a Fully-Spatial Reconfigurable Architecture 113
References 117

12 Design Space Exploration for MAHA Framework 119
12.1 Benchmarks and Simulation Setup 119
12.2 MLB Architecture Exploration 120
References 124

13 Preferential Memory Design for MBC Frameworks 125
13.1 Case Study—I: SRAM Array 125
13.2 Case Study—II: STTRAM Array 130
13.3 Optimization in Application Mapping 133
References 136

14 PLA Based Application Mapping in MBC 137
14.1 Function Representation as PLA 137
14.2 Function Representation Using CAM Based MBC Framework 138
14.3 Power and Performance of a CAM Based MBC Framework 141
References 143

Part VI Off-Chip Hardware Acceleration Using MBC

15 Background and Motivation 147
15.1 Von-Neumann Bottleneck 147
15.2 Mitigating Von Neumann Bottleneck Through In-Memory Computing 152
References 156

16 Off-Chip MAHA Using NAND Flash Technology 157
16.1 Overview of Current Flash Organization 157
16.2 Modifications to Flash Array Organization 160
16.3 ECC Computation and Modes of Operation 163
References 163

17 Improvement in Energy-Efficiency with Off-Chip MAHA 165
17.1 Design Space Exploration for Off-Chip *MAHA* 165
17.2 Energy and Performance for Mapped Applications 169
17.3 Hardware Emulation Based Validation 173
References 176

Part VII Improving Reliability of Operations in MBC

18 Mitigating the Effect of Parametric Failures in MBC Frameworks 179
18.1 Effect of Parameter Variations on QoS 179
18.2 A Variation-Aware Preferential Mapping Approach 183
18.3 Preferential Memory Design 188
References 193

19	**Mitigating the Effect of Runtime-Failures in MBC Frameworks**	195
	19.1 Impact of Parameter Variation on Runtime Failures	195
	19.2 Reliability-Driven ECC Allocation for Multiple Bit Error Resilience	196
	19.3 Experimental Results	201
	References	207
20	**Summary**	209

Acronyms

CAM	Content-addressable memory
CPU	Central processing unit
CDFG	Control data flow graph
CGRA	Coarse-grained reconfigurable array
CMOS	Complementary metal-oxide semiconductor
DRAM	Dynamic random-access memory
EDP	Energy-delay product
FPGA	Field-programmable gate array
FeRAM	Ferro-electric random access memory
GPP	General purpose processor
GPU	Graphics processing unit
LUT	Lookup table
MBC	Memory-based computing
MLB	Memory logic block
MAHA	Malleable hardware accelerator
MRAM	Magnetoresistive random access memory
NVM	Non-volatile memory
PE	Processing element
PI	Programmable interconnect
PCM	Phase change memory
PLA	Programmable logic array
PTM	Predictive technology model
RRAM	Resistive random access memory
SRAM	Static random-access memory
STTM	Spin-transfer torque memory
VLSI	Very-large-scale integration

Part I
Introduction

Chapter 1
Challenges in Computing for Nanoscale Technologies

Abstract In the era of nanoscale technologies, computing systems are faced with a set of common challenges, namely energy-efficiency and reliability. This challenge applies to varieties of computing systems namely software programmable general-purpose processors, hardware reconfigurable computing frameworks and application-specific custom hardware. As demonstrated by numerous prior works from both industry and academia, an energy-efficient and reliable VLSI system in nanoscale technologies is actually an ensemble of the different computing approaches mentioned above. In this chapter, we discuss each of the challenges in detail, particularly in the context of the emerging workloads for modern VLSI systems.

1.1 End of Dennard Scaling

In his seminal paper on metal-oxide semiconductor (MOS) device scaling in 1974, Robert Dennard showed that when voltages are scaled along with all dimensions, a device's electric fields remain constant, and most device characteristics are preserved [1]. Following Dennard's scaling theory, chip makers achieved a triple benefit: First, devices became smaller in both the x and y dimensions, allowing for $1/\alpha^2$ more transistors in the same area when scaled down by α, where $\alpha < 1$. Second, capacitance scaled down by α, thus, the charge Q that must be removed to change a node's state scaled down by α^2. The current also scaled down by α, so the gate delay D decreased by α. Finally, because energy is equal to CV^2, energy decreased by α^3. Most importantly, following Dennard scaling maintained constant power density. Thus, scaling alone was able to bring about significant growth in computing performance at constant power profiles.

Taking advantage of Dennard's scaling, designers continued to aggressively design their circuits by incorporating faster memories, adders, and latches, and created deeper pipelines that increased clock frequency. During this period, computer designers wisely converted a "free" resource (extra power) into increased

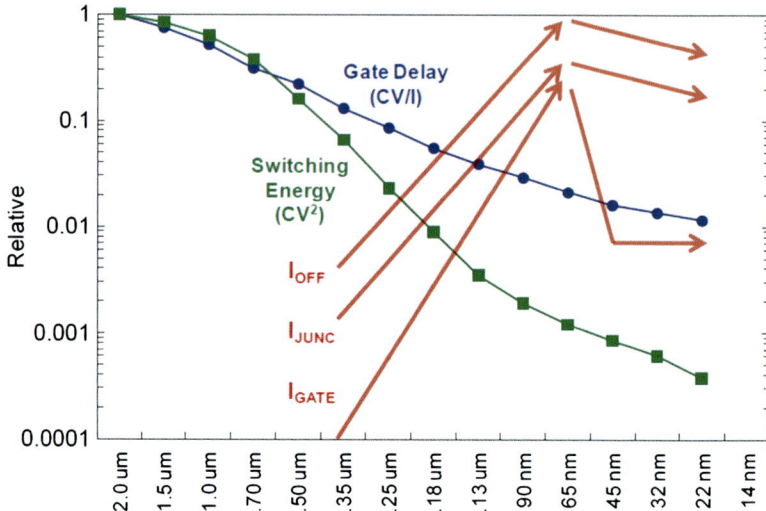

Fig. 1.1 Scaling trends showing transistor performance and switching energy across technology generations [2]

performance, which everyone wanted. In the early 2000s, however, high performance designs reached a point at which they were hard to air-cool within cost and acoustic limits. Moreover, the laptop and mobile device markets which are battery constrained and have even lower cooling limits were growing rapidly. Thus, most designs had become power constrained. At around the same time, technology scaling began to change as well. Up until the 130-nm node, supply voltage (V_{dd}) had scaled with channel length. But at the 90-nm node, V_{dd} scaling slowed dramatically, faced with exponentially increasing leakage power costs, if the threshold voltage V_{th} is also scaled with V_{dd}. As a result, V_{dd} scaling has stopped and it is still around 1 V for the 45-nm node. With constant voltages, energy now scales with α rather than α^3, and as we continue to put $1/\alpha^2$ more transistors on a die, we are facing potentially dramatic increases in power densities unless we decrease the average number of gate switches per second. Although decreasing frequencies would accomplish this goal, it isn't a good solution, because it sacrifices performance. In the power-constrained, post-Dennard era, creating energy-efficient designs is critical. Continually increasing performance in this new era requires lower energy per operation, because the product of operations per second (performance) and energy per operation is power, which is constrained. Figure 1.1 shows the transistor scaling trends for Intel. As evident from the figure, drastic improvements in either performance or energy can no longer be expected from technology scaling.

In the past, the push for ever better performance has seen designs creep up to the steep part of this energy-performance tradeoff curve. However, as Pollack's rule states [3], the performance for a single CPU system only increases as $\sqrt{(r)}$, where r represents the number of transistors in the design. This has forced designers to reevaluate the situation, and this is precisely what initiated the move to multicore systems. Of course, this approach sacrifices single-threaded performance, and it also assumes that the application is parallel which isn't always true. However, we cannot rely on parallelism to save us in the long term for two reasons. First, as Amdahl noted in 1965 [4], with extensive parallelization, serial code and communication bottlenecks rapidly begin to dominate execution time. Thus, the marginal energy cost of increasing performance through parallelism increases with the number of processors, and will start increasing the overall energy per operation. The second issue is that parallelism itself doesn't intrinsically lower energy per operation; lower energy is possible only if sacrificing performance also yields a lower energy point in the energy-performance space. A widespread initiative across academia and industry has therefore been undertaken to find a solution to sustaining the performance improvement with technology scaling without violating the power constraint [6].

1.2 Sustaining Reliability of Operation

Even with constant device failure rates, technology scaling makes sustaining system level reliability substantially more difficult as the number of transistors that are integrated on-die increases. At the architecture level, the compute system must have the ability to identify system issues such as failures, and should to be able to operate through and optimize performance with available system resources. As transistor dimensions scale, their operation are more prone to: (i) increased electric fields; (ii) shrinking $V_{max} - V_{min}$ window; (iii) thermo-mechanical limitations and (iv) soft-errors. These physical causes often exhibit at the system-level as the following effects: (i) *Hard Failures*; (ii) *Transient Failures* and (iii) *Parametric degradation*. Table 1.1 shows the trends for each of these effects as the technology is scaled. In order to counteract these failure mechanisms, both circuit

Table 1.1 Trends in the nature of CMOS failure at nanoscale technologies [5]

Fail Type	Trend	Solution Space
Hard Fails	Flat	Continued Process Improvements
		Architectural features (e.g. spare rows)
Parametric degradation	Increasing	Continued Process Improvements
		Guardbands
		Architectural Fault Tolerance
Transient Noise	Increasing	Continued Process Improvements
		Guardbands
		Architectural Fault Tolerance
Radiation induced Noise	Increasing to flat	Improved estimation methodology
		Hardening of critical elements
		Architectural fault tolerance

and architecture-level fault tolerance schemes have been proposed. Among the popular architecture/system-level schemes, (i) error-correction coding; (ii), sparing and replication and (iii) dynamic error-detection and rollback have emerged as the popular design choices [6]. Note that these resiliency techniques are extremely popular for protection of data stored in the memory [7, 8]. The primary reason being the overhead due to these resiliency techniques are easily amortized over the size of the memory being protected. On the other hand, the random logic or custom datapath are difficult to protect from manufacturing defects and runtime failures.

1.3 Addressing Energy-Efficiency for Data-Intensive Workloads

In scaled technologies, instead of simply looking at the energy and reliability challenges for micro-processors or SoCs on-die, researchers have shifted their attention to system-level energy and reliability management. The primary reason being for modern workloads, data-movement between off-chip memory (primary or secondary) and on-chip processing units, has emerged as the major bottleneck to system-wide energy scaling. This concern has been emphasized in the technical report on Exascale Computing as published from DARPA in 2008 [6]. With technology scaling, on-die memory arrays and logic scale in performance and power. The same is however not true for off-chip links which do not scale at the same rate. On-chip global interconnects are much worse as shown in Fig. 1.2. It is therefore a general consensus that interconnect is emerging as the serious bottleneck and will soon dominates the delay and power in both logic and memory systems [10]. In addition, memory latency and bandwidth has always been a bottleneck compared to logic-only blocks. Conventionally architecture-innovations such as out-

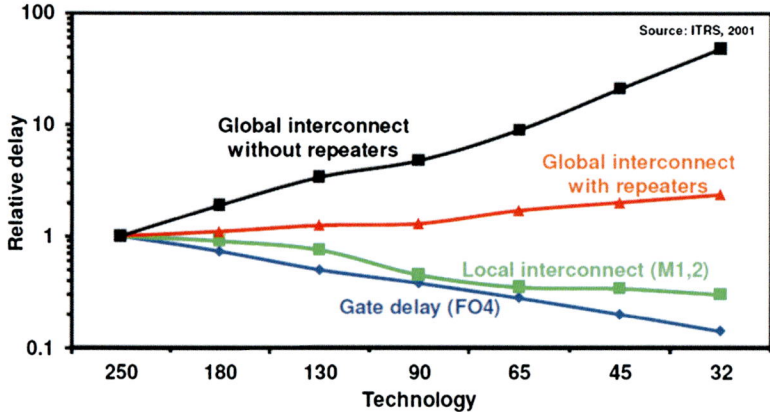

Fig. 1.2 Performance scaling trends for on-chip interconnect with technology generations [9]

1.3 Addressing Energy-Efficiency for Data-Intensive Workloads

Table 1.2 Taper in operands per unit energy with access to higher levels in the memory hierarchy

Path	Energy(pJ) Total	Transport	Access	% Transport	Operands per pJ
Register File	5	0.0	5	0.0	1.436
L1 Hit	39	0.0	39	0.0	0.26
L1 miss, L2 Hit	385	167.3	217.7	43.4%	0.033
L1 miss, L3 Hit	1991.7	1774	217.7	89.6%	0.006
Cache Miss, On-module DRAM access	1380	1204.7	175.26	87.3%	0.003
On-module DRAM miss Off-module DRAM access	13819	13791.4	27.63	99.8%	0.000

of-order operation had hid the memory and interconnect latency. But as correctly pointed in [6], "it is possible to hide the delay but not the energy". Thus there exists a requirement to appropriately the schedule the tasks of an application in both space and time and manage communication across the storage hierarchy explicitly.

The situation is exacerbated by the rise of workloads which are data-intensive, such as multimedia processing, data-mining and others. For such workloads, *Von-Neumann* bottleneck has emerged as a key-limiter to both performance and energy scaling. The energy overhead is accessing off-chip memory is easily observed in Table 1.2. As the memory-hierarchy level from which data is being fetched to on-chip compute modules increases, the energy spent increases by order of magnitude. This means for the same energy budget, less and less data can be brought from memories higher up in the hierarchy, such as, primary memory like DRAM and secondary memory like Flash. A detailed study in [11] shows that in a large scale multiprocessor system, a significant percentage (75%) of energy is spent in data movement. As summarized in Table 1.2, only a small fraction of energy is actually spent in reading the data from different levels of the memory hierarchy. Most of it is spent in data transport across the hierarchy. This reduces the computational efficiency for real applications to less than 5%. It is therefore vital that data-movement across memory hierarchies are well managed and off-chip accessed are minimized as much as possible.

1.3.1 Poor Scaling of Off-Chip Bandwidth

Computing systems such as FPGA and GPU often demand high off-chip bandwidth. The requirement is typical for the parallel applications which are mapped to these non-Von-Neumann computing systems. Although these frameworks have been found to offer 10–1,000× improvement in energy-efficiency compared to a single general-purpose processor, their performance finally limited by off-chip memory bandwidth and energy by off-chip memory access energy. Off-chip bandwidth is not scaling at the same rate on-chip logic density. As pointed out in [12], while

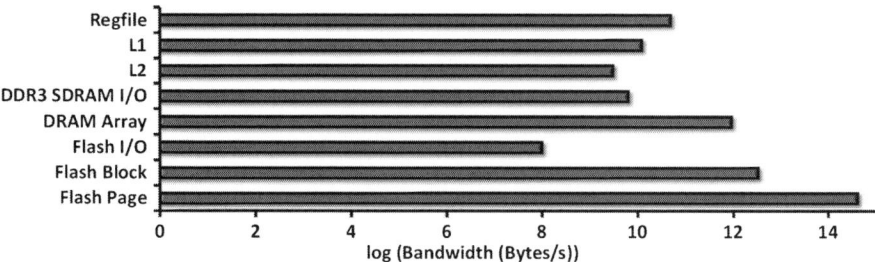

Fig. 1.3 A comparison of the bandwidths at different levels of the memory hierarchy in a modern computing system

the logic density increases by a factor of 16× from 2011 to 2024, the relative bandwidth only increase by 1.4× over the same period of time. As a result for data-intensive applications, both GPU and FPGA will soon be limited by off-chip data bandwidth [12]. The scenario is however completely opposite if a computing framework that resides inside the memory is considered. As shown in Fig. 1.3, the data highest bandwidth can be achieved inside the commercially available non-volatile memory technologies such as Flash. The large bandwidth is mainly due to the wide (2KB) page read performed in each of the individual Flash blocks (total number of blocks is typically 4,096) at $\sim$ 15 ns. The bandwidth sharply falls as data is transferred out from the Flash at an I/O interface which is only 16-bits wide. The same scenario holds for data bandwidth inside the conventional DRAM. Note that the Intelligent RAM (IRAM) [13] was implemented to leverage on this observation and mitigate the Von-Neumann bottleneck. From Fig. 1.3, we note that compared to internal DRAM bandwidth, the bandwidth between the main memory and the processor is order of magnitude less. Thus to sustain the requirement for energy-efficient execution of parallel data-intensive workloads, researchers must investigate the feasibility of novel computing architectures which exploit computing in off-chip non-volatile memories.

References

1. O. Shacham et al., "Rethinking Digital Design: Why Design Must Change", in *Intl. Symp. Microarchitecture*, 2010
2. M. Bohr, "Silicon Technology Leadership for the Mobility Era". http://www.intel.com/content/dam/www/public/us/en/documents/presentation/silicon-technology-leadership-presentation.pdf
3. F. Pollack, "New Microarchitecture Challenges in the Coming Generations of CMOS Process Technologies", in *Intl. Symp. on Microarchitecture*, 1999
4. G.M. Amdahl, "Validity of the Single Processor Approach to Achieving Large-Scale Computing Capabilities", in *AFIPS*, 1967
5. X. Vera-Raimat, "DFx for Terascale Reliable Processors". http://www.cse.psu.edu/~yuanxie/ISCA10-WTAI/Reliability-Vera-Intel.pdf

References

6. P. Kogge et al., "ExaScale Computing Study: Technology Challenges in Achieving Exascale Systems". http://www.cse.nd.edu/Reports/2008/TR-2008-13.pdf
7. S. Paul, F. Cai, X. Zhang, S. Bhunia, "Reliability-driven ECC allocation for multiple bit error resilience in processor cache". IEEE Trans. Comput. **60**(1), 20–34 (2011)
8. S. Mukhopadhyay, H. Mahmoodi, K. Roy, "Modeling of failure probability and statistical design of SRAM array for yield enhancement in nanoscaled CMOS". IEEE Trans. Comput. Aided Des. Integrat. Circ. Syst. **24**(12), 1859–1880 (2005)
9. "International Technology Roadmap for Semiconductors 2001 Edition : Interconnect". http://www.itrs.net/Links/2001ITRS/Interconnect.pdf
10. [Online], "ITRS 2007 : Interconnect". http://www.itrs.net/links/2007itrs/home2007.htm
11. [Online], "Future processing : Extreme scale". http://www.hpcuserforum.com/EU/downloads/DARPAExtremscaleHarrod_ACS_Bri%ef_01709.pdf
12. E.S. Chung, P.A. Milder, J.C. Hoe, K. Mai, "Single-Chip Heterogeneous Computing: Does the Future Include Custom Logic, FPGAs, and GPGPUs? in *Intl. Symp. on Microarchitecture*, 2010
13. [Online], "The Berkeley Intelligent RAM (IRAM) Project". http://iram.cs.berkeley.edu/

Chapter 2
A Survey of Computing Architectures

Abstract In this chapter, we review the computing architectures popularly used in commercial systems. This includes general-purpose processors, graphics processing units and FPGAs. Motivated by the fact that hardware reconfigurable frameworks like FPGAs can significantly benefit energy-efficiency for compute-intensive workloads, we then discuss some well-known hardware reconfigurable architectures. Finally we discuss the scaling challenges for fully-spatial reconfigurable architectures such as FPGAs.

2.1 Von-Neumann and Non-Von-Neumann Architectures

General purpose micro-processors are classical examples of Von-Neumann machines. Data which needs to be processed is brought to the compute pipeline through a series of memory copy operations. Such architecture is typically characterized by clear separation of compute logic and memory storage. Another characteristic of such systems is that they are temporal architectures, with one operation being executed in one clock cycle and dependent operations being executed across multiple clock cycles. Although super-scalar machines with parallel execution units can substantially improve performance, the degree of parallelism is still limited compared to those offered by non-Von-Neumann architectures such as FPGAs and GPUs. These architectures typically incorporate thousands of processing elements supporting similar number of threads. While processing elements are fine-grained arithmetic units for FPGA, execution units for GPUs are designed to support complex floating-point operations. They are classified as non-Von-Neumann since compute and memory operate in close association. Instructions are typically downloaded into these hardware architectures, allowing them to be pre-configured prior to actual operation. This is contrary to Von-Neumann architectures where both data and instruction are fetched at the same time. With the end of Dennard scaling, conventional Von-Neumann architectures have been shown to be insufficient in satisfying the growing energy requirements for modern data-intensive workloads. Heterogenous architectures with Von-Neumann components for single-thread efficiency and non-Von-Neumann for efficiency in multi-threaded workloads has therefore been commercialized.

Researchers and developers are constantly enlarging the application space and improving the performance of computing machinery. Some examples are weather prediction and nuclear explosion simulation in scientific computing, DNA sequence matching in bioinformatics, high-speed media applications over mobile networks, high speed switching and routing and so forth [1]. Applications from these diverse domains have been traditionally computed on a software reconfigurable platform such as a general-purpose processor (GPP). However, the energy-efficiency measured in terms of performance per watt achieved from a GPP has not found to scale with technology scaling [2]. ASICs have always been found to offer the best performance per watt among all the computing platforms [3]. The downside is however, such a platform is committed to one or more applications during fabrication and therefore possesses limited flexibility in executing arbitrary applications. Reconfigurable computing is a novel computing paradigm which combines the flexibility of software with the high performance of application specific hardware, bridges the gap between general-purpose processors and application-specific systems, and enables higher productivity and shorter time to market [4]. Figure 2.1 illustrates the same idea. As we note from Fig. 2.1, by definition, ASICs have a fixed functionality and only operate on input data. Processors are software reconfigurable and the application is defined in terms of instructions. Systems such as FPGAs are however hardware reconfigurable and the input application is defined in terms of configuration bits which must be loaded into the reconfigurable framework before it executes an application.

2.2 Reconfigurable Computing: State of the Art

In this section, we focus our attention on hardware reconfigurable frameworks such as FPGAs. These systems are capable of mapping hardware descriptions of arbitrary applications and extremely efficient in their operations. In this section, we review prior works on hardware reconfigurable architectures and identify the major challenges confronting reconfigurable hardware design nanoscale technologies. We also discuss possible avenues to mitigate those challenges and extend our discussion to how non-Von-Neumann architectures can be more effective than their Von-Neumann counterpart in mitigating these challenges.

2.2.1 Taxonomy for Reconfigurable Systems

A reconfigurable computing system consists of an array of *reconfigurable* processing elements (PEs), the behavior for each of which as well as their connectivity can be customized based on the application which is to be mapped

2.2 Reconfigurable Computing: State of the Art

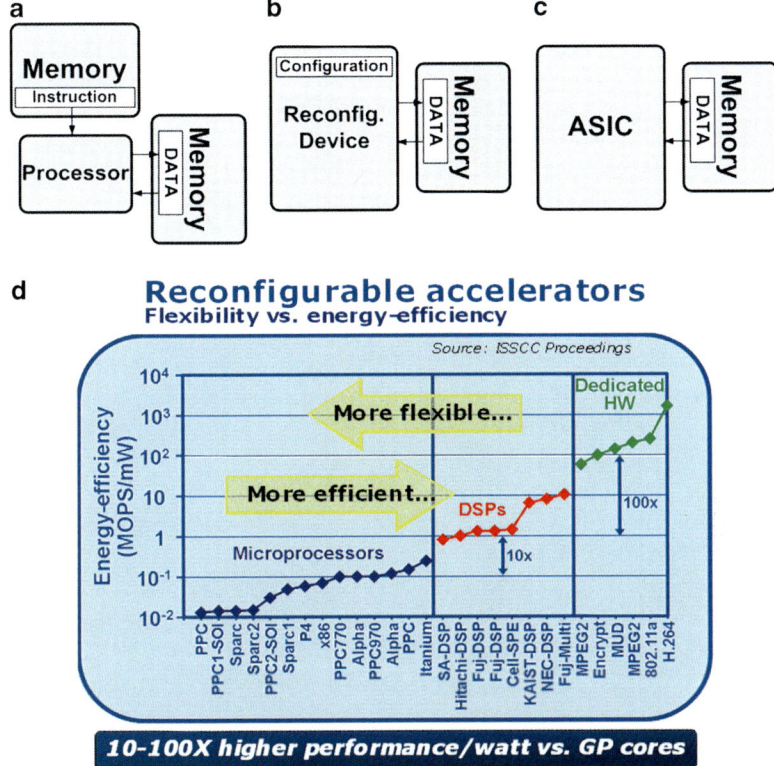

Fig. 2.1 (a) A Von-Neumann computing framework like GPP. (b) A hardware-reconfigurable framework like FPGA. (c) An application-specific hardware. (d) Reconfigurable accelerators sacrifice flexibility but provide higher-energy efficiency compared to GPP

to the reconfigurable system. The programmable interconnect which connect these processing elements together is responsible for steering the inputs and outputs as well as intermediate results to correct processing elements. A reconfigurable architecture with adequate interconnect and compute resources greatly improves the flexibility, resource utilization and therefore provides superior performance. In addition to the reconfigurable frameworks which are commercially available, a large number of reconfigurable architectures have been proposed till date. These can be distinguished from one another by specifying some or all of the following characteristics for each of these reconfigurable frameworks.

1. **Granularity:** The granularity of a reconfigurable architecture is defined as the width of the smallest processing element. Based on the granularity, reconfigurable systems can be classified into *fine-grained* and *coarse-grained* reconfigurable systems. In fine-grained systems, each PE primarily consists of logic gates, flip flops and lookup-tables. They operate at bit level, implementing a boolean function of a finite-state machine. On the other hand PEs in

coarse-grained architectures contain complete functional units, like ALUs and/or multipliers that operate on multi-bit words. Coarse-grained reconfigurable frameworks have been shown to significantly reduce the PI overhead for the fine-grained reconfigurable framework and reduce the complexity for Place and Route (P&R) problems [5]. However, the major drawback of coarse-grained reconfigurable architectures is that the optimal granularity of the reconfigurable resources varies from one design to another. Thus when the application data does not match with that of the logic in a coarse grained framework, performance decreases [6]. Moreover, the specialized coarse-grain functional units and very structured communication resources makes application mapping less flexible for these architectures [6]. Many modern reconfigurable systems including FPGAs therefore combine the benefits of both *fine-grained* and *coarse-grained* system and are known as *mixed granular* reconfigurable systems.

2. **Depth of Programmability:** This pertains to the number of configuration programs or *contexts* that may be stored inside each PE. For *multi-context* PEs, contexts for several programs reside concurrently in the system. This enables the execution of different tasks by simply changing the operating context wihtout having to reload the configuration program.

3. **Reconfigurability:** In a reconfigurable system, reconfiguration is the process of reloading configuration programs (contexts). The process can be either *static* (execution is interrupted) or *dynamic* (in parallel to the execution). Dynamic reconfiguration is more relevant for multi-context PEs. This feature significantly reduces the overhead for reconfiguration.

4. **Interface:** Reconfigurable computing systems can work as a stand alone computing system as well as in conjunction with a system's host processor. The reconfigurable system has a *remote* interface if it is not on the same die as the host processor and communicates over some I/O bus, e.g. PCI bus. The interface is *local* if the reconfigurable system and the host processor reside on the same die. The communication is similar to a co-processor model and is over a fast on-chip bus. A reconfigurable system is called *tightly coupled* if it is placed inside the processor. The instruction unit issues instruction to the reconfigurable unit as if it is a standard functional unit of the processor. Possible interfaces between the reconfigurable functional unit (RFU) and the host processor is shown in Fig. 2.2.

5. **Application Domain:** Different reconfigurable frameworks cater to different application domains. The suitability towards a particular application domain is mainly determined by the granularity of the framework and the computational model which may be single instruction multiple data (SIMD) or multiple instruction multiple data (MIMD) models.

Table 2.1 shows the classification of some of the popular reconfigurable frameworks based on the above taxonomy.

Fig. 2.2 Possible couplings between the CPU core and a reconfigurable functional unit

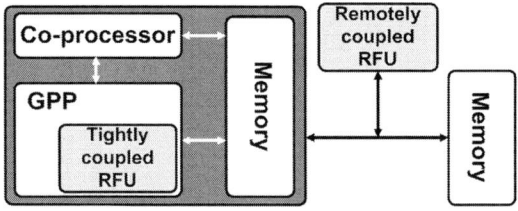

Table 2.1 Taxonomy of modern reconfigurable frameworks

	Granularity	Programmability	Reconfiguration	Interface	Appl. Domain
FPGA	Mixed	Single	Static	Local/remote	General Purpose
DPGA	Fine	Multiple	Dynamic	Remote	Bit-level computations
RaPiD	Coarse	Single	Static	ND*	Systolic nature, easily pipelined
Chimaera	Fine	Single	Static	Tightly Coupled	Bit-level computations
Matrix	Coarse	Multiple	Dynamic	ND*	ND*
Piperench	Mixed	Multiple	Dynamic	Remote	DSP
Morphosys	Coarse	Multiple	Dynamic	Tightly coupled	Image Proc.

ND* - Not defined

2.2.2 A Brief Survey of Reconfigurable Frameworks

Here we briefly describe some of the reconfigurable architectures proposed till date.

1. **Field Programmable Gate Arrays (FPGAs):** The most popular FPGA arrangement today is the *island-style* architecture [7]. As seen from Fig. 2.3, it consists of computational resources which are divided into small islands of logic blocks which are surrounded by a sea of interconnect wires and programmable communication resources. Each logic block can generally implement any function of N-inputs 1-output through the use of Look-up tables (LUTs). LUTs are simply small memories which allow the inputs to the logic block to address a particular location in a read-only memory. By populating the LUTs appropriately, the user may map any N-input 1-output function to these LUTs. FPGAs also offer the opportunity to implement sequential logic by providing an optional flip-flop at the output of the LUTs. The LUT/flip-flop pair is sometimes referred to as the basic logic element (BLE).

    The communication resources required for an island style FPGA can be divided into three components: (i) *channels* are groups of individual wires that are locally organized into bundles by their physical location. Figure 2.3a shows a channel width of 4. (ii) *Connection boxes* are responsible for managing the movement of data into and out of the channel. They control the wires which become logic block input or primary output or become logic block output or primary input. (iii) *Switch boxes* are responsible for connecting wires in different channels together. When synthesizing arbitrary netlists into FPGA,

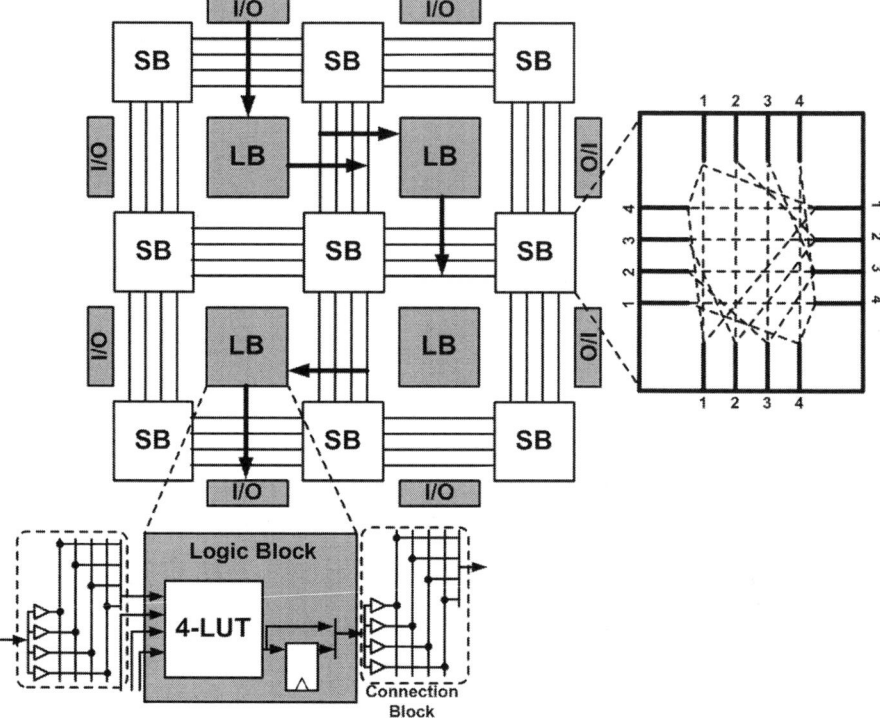

Fig. 2.3 Conventional island-style FPGA

the mapping tool maps the input netlist to LUTs and flip flops and binds them to the logic block. These blocks are then placed and routed using the programmable interconnects.

In order to bridge the large area, energy and performance gap that exists between pure LUT based mapping and custom logic implementation for some functions, FPGA vendors have integrated more components on the FPGAs, such as general purpose processors, custom adders and multipliers for digital signal processing and block RAM modules. Moreover, in addition to single length wires as shown in Fig. 2.3a, they have interconnect wires that span multiple logic blocks and improve the speed for long distance communication. FPGAs may also include *dedicated carry chains*. Lastly, in order to improve energy and performance, modern FPGAs cluster together multiple BLEs within a single logic block/cluster, which allows BLEs within a cluster to communicate with each other without using external routing wires.

2. **Dynamically Programmable Gate Arrays (DPGAs):** DPGA is a programmable array which allows the strategic reuse of limited resources. In so doing, DPGAs promise greater capacity, and in some cases higher performance, than conventional programmable device architectures where all array resources

2.2 Reconfigurable Computing: State of the Art

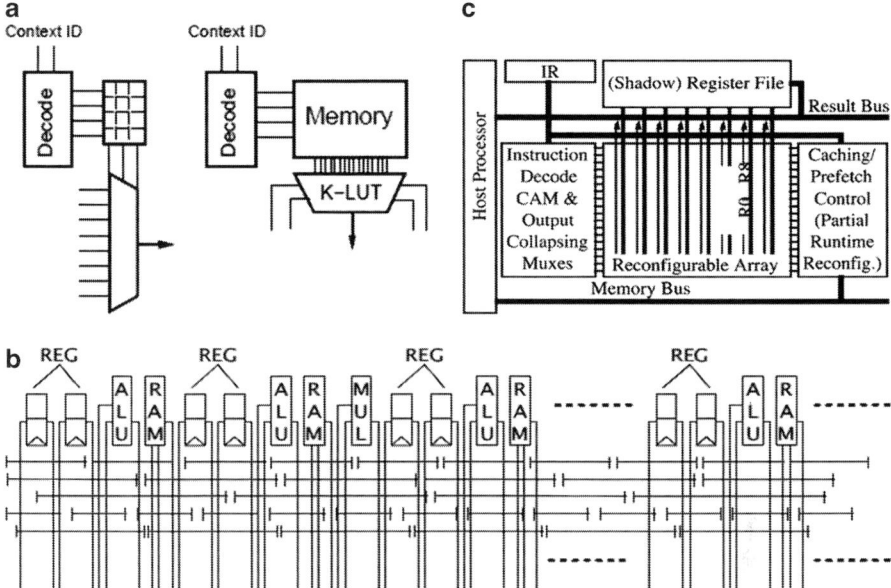

Fig. 2.4 (**a**) Context memory at each logic block and routing switch in a DPGA architecture [8]. (**b**) RaPiD—a linear reconfigurable array [9]. (**c**) Chimaera reconfigurable architecture [10]

are dedicated to a single function for an entire operational epoch. This is achieved by building a multicontext FPGA, which holds several configuration for each gate and switch as shown in Fig. 2.4a. A broadcast context identifier tells each primitive which configuration to select at any given point in time. Through techniques such as temporal pipelining, utility functions, multiple function accommodation and state dependent logic, DPGAs have been proven to effectively improve resource utilization for a variety of applications [8].

3. **Reconfigurable Pipelined Datapath (RaPiD):** is a linear array of PEs, which is configured to form a linear computational pipeline, as shown in Fig. 2.4b. Each PE comprises an integer multiplier, three integer ALUs, some general-purpose data registers, and three small RAMs. A typical RaPiD contains up to 32 such PEs. These PEs are interconnected using a set of segmented buses that run the length of the linear array. The buses are segmented into different lengths and placed in different tracks. The framework distributes computations over the pipeline stages and is amenable to mapping of regular streaming applications [9].
4. **Chimaera:** It integrates a reconfigurable logic into the host processor (refer to Fig. 2.4c). Sequences of instructions are mapped to one or more lines of the reconfigurable array. In some cases, sequences of instructions are collapsed and then mapped. Reconfigurable operations (RFUOPs) are extracted using the Chimaera compiler. Chimaera is a fine grained architecture which supports multi-operand instruction. It does not however have any state elements and uses

the *shadow register file* of the host processor for temporary storage. Because the reconfigurable operations are extracted at compile time, the Chimaera architecture cannot span multiple basic blocks. Also the mapped operations are still dependant in the reconfigurable array, so the operations are not really collapsed.

5. **MATRIX:** As shown in Fig. 2.5a is a coarse-grained reconfigurable array which supports configurable instruction distribution. MATRIX is an improvement over the DPGA concept. Rather than separate the resources for instruction storage and distribution from the resources for data storage and computation and dedicate silicon resources to them at fabrication time, the MATRIX architecture unifies these resources. Once unified, traditional instruction and control resources are decomposed along with computing resources and can be deployed in an application-specific manner. The adaptability is made possible by a multi-level configuration scheme, a unified configurable network supporting both datapaths and instruction distribution, and a coarse-grained building block which can serve as an instruction store, a memory element, or a computational element. At 0.5μ process point, MATRIX was demonstrated to achieve 10Gops, running at 100Mhz.

6. **Piperench:** It functions as a remotely-coupled reconfigurable co-processor for pipelined applications (refer to Fig. 2.5b). PipeRench contains a set of physical pipeline stages called stripes. Each stripe has an interconnection network and a set of PEs. Each PE contains an arithmetic logic unit and a pass register file. Each ALU contains lookup tables (LUTs) and extra circuitry for carry chains, zero detection, and so on. Through the interconnection network, PEs can access operands from registered outputs of the previous stripe, as well as registered or unregistered outputs of the other PEs in the same stripe. Moreover, the PEs access global I/O buses. Piperench overcomes the cost issues of implementing a circuit using FPGA cells by implementing a large circuit on a small number of FPGA cells through a technique known as *hardware virtualization*. Piperench is suitable for stream-based media applications and regular fine-grained computations.

7. **Morphosys:** Morphosys as illustrated in Fig. 2.6 is a coarse-grained reconfigurable array which is tightly coupled with a general purpose processor for accelerating word-level compute-intensive applications. It is targeted at high-throughput and data-parallel applications. Morphosys incorporates a 2-D array of reconfigurable cells organized in a SIMD fashion and also a high bandwidth memory interface that consists of a streaming buffer to handle data transfers between external memory and the Morphosys array. The suitability of the Morphosys framework has been demonstrated for target application domains such as video compression, data encryption and target recognition.

In addition to the above frameworks, a large number of other stand-alone reconfigurable frameworks [13–21] can be found in literature. While some of these are for general purpose applications [18–20], others are targeted towards a specific application domain [16, 17, 21].

8. **ASIPs and RISPs:** Reconfigurable frameworks often work in conjunction with a host processor to map data-parallel and compute-intensive kernels which

2.2 Reconfigurable Computing: State of the Art

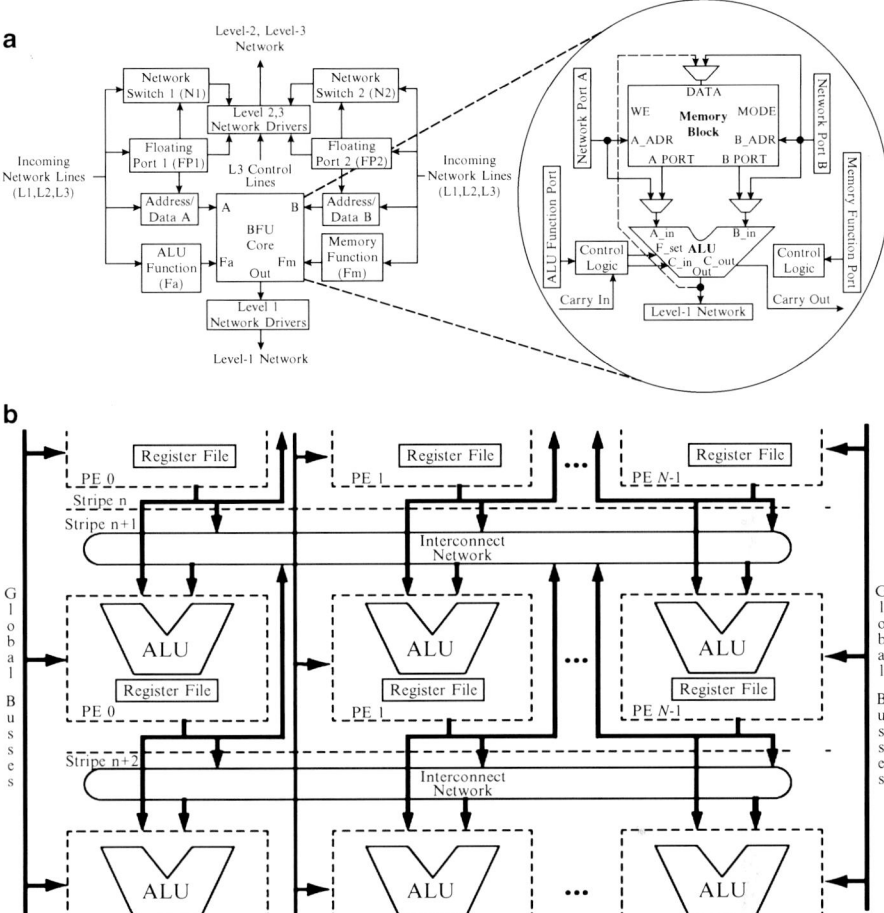

Fig. 2.5 (a) Matrix reconfigurable architecture with configurable instruction distribution and deployable resources [11]. (b) Piperench—an attached reconfigurable co-processor for pipelined applications [12]

become performance and energy bottleneck in simple software based execution [5]. Reconfigurable solutions in processor systems can be classified into two categories: (i) static and (ii) dynamic. The static solution otherwise known as application specific instruction set processors (ASIPs) are statically configured during their design and cannot be configured at runtime. The idea is to first identify the application domain for which a processor will primarily be used. Next common characteristics are extracted from applications pertaining to that application domain. Following this specialized hardware units are incorporated into the processor at design time and the processor instruction set is altered to support the operation for this specialized unit. During runtime, this hardware unit

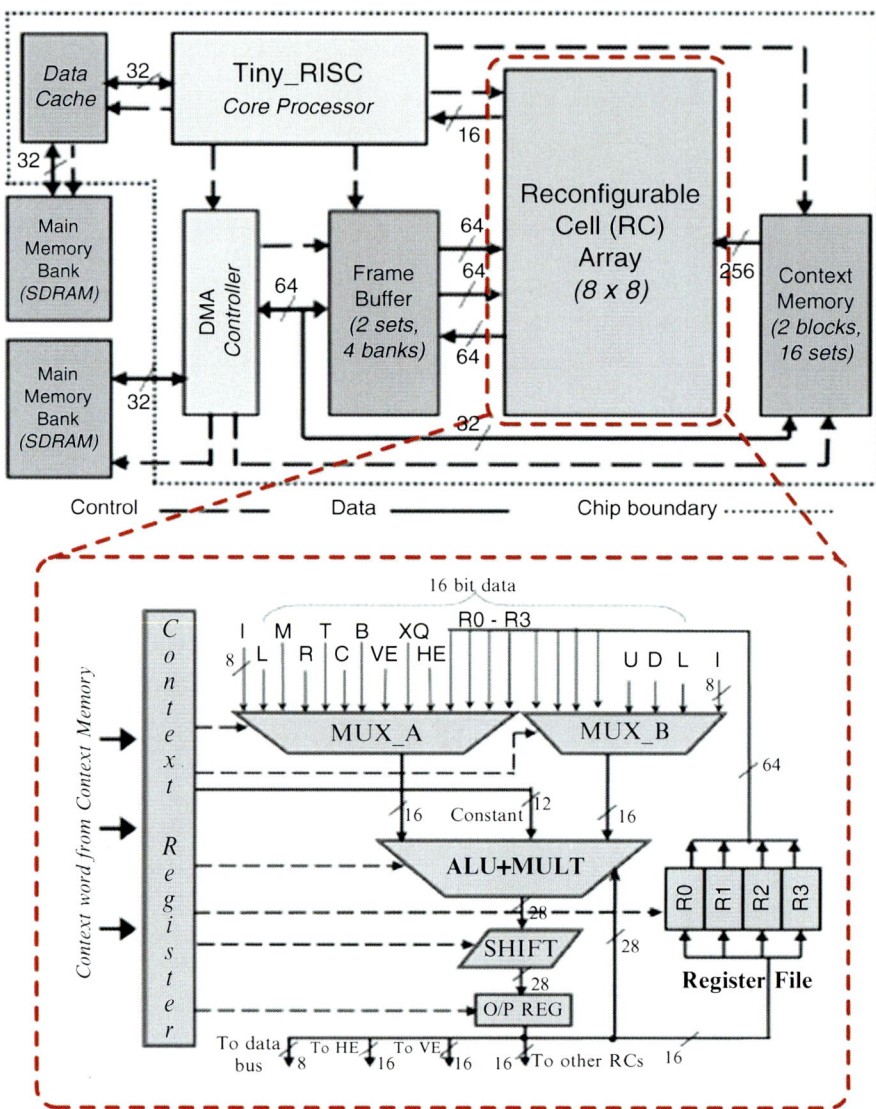

Fig. 2.6 Morphosys—An integrated reconfigurable array for data-parallel and compute-intensive applications [24]

is used to accelerate domain specific instruction calls which would otherwise be costly to execute with the primitive instruction set. Tensilica's Xtensa customizable processors [22] and Stretch's S7000 software configurable processors [23] are examples of ASIPs.

In contrast to ASIPs which are statically designed, it is possible to incorporate a RFU inside the processor so that the system may dynamically adapt to computing demands from different application domains. Such architectures are referred to as reconfigurable instruction set processors (RISPs). A RISP has more adaptability compared to an ASIP. There are however more hardware and software aspects that one needs to consider during the design of such a reconfigurable processor. These are: (i) coupling of the processor and the reconfigurable logic, (ii) configuration, (iii) instruction coding and scheduling, (iv) granularity, (v) hardware cache and (vi) reconfigurability. A detailed survey of reconfigurable processor architectures and their design considerations can be obtained in [5, 25].

2.3 Challenges in Reconfigurable Hardware Design

As technology scales reconfigurable systems are confronted by a number of challenges at all levels of design abstraction, namely, circuit, architecture and software. The common challenges which applies to all three are (i) minimizing the power for reconfigurable frameworks, (ii) improving the performance of the reconfigurable framework and (iii) tolerate defects. Here we try to summarize the solutions which have been proposed to address these challenges.

2.3.1 Circuit

1. **Power:** In nanoscale technologies, power has emerged as the primary constraint to the design of compute systems [26]. Since reconfigurable systems are being used more and more in mainstream computing, there arises a strong demand for reducing the power requirement for FPGAs and similar other reconfigurable systems. At the same technology node, a FPGA has been found to consume $12\times$ the dynamic power compared to an ASIC implementing the same application [27]. The major contribution to power comes from the programmable interconnect which has been found to contribute 60% [28] to the total dynamic power and 55% [29] to the total leakage power in the FPGA platform. Circuit level techniques for power reduction traditionally incorporate dual-V_{dd} and/or dual-V_t fabrics in the same FPGA device. Designs are mapped to a combination of computing elements with high and low V_{dd} and V_t depending on the performance constraint for dynamic and leakage power reduction [30]. Leveraging on this dual V_{dd} and V_t assignment, low power designs of the logic block [31] and routing switches [32] have been previously proposed. Region based power gating [33] as well as fine-grained power gating [34] have also been proposed for power reduction in FPGA frameworks.

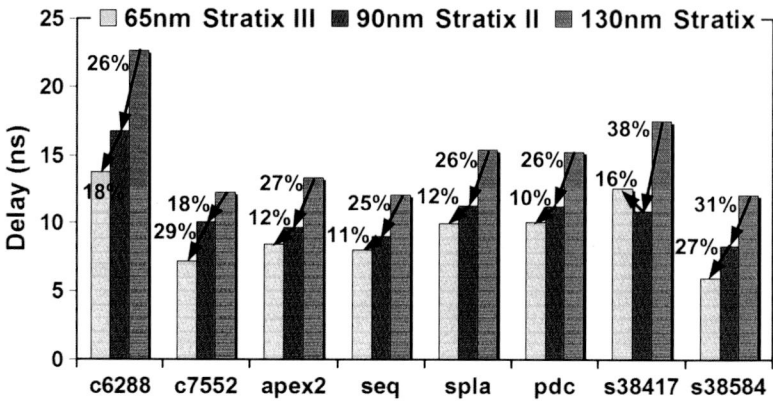

Fig. 2.7 Performance scaling trends for different benchmarks mapped to Altera Stratix FPGA family

2. **Performance:** Like power, interconnect is also the performance limiter in case of conventional FPGA frameworks. For modern FPGAs, interconnect delay has been shown to contribute 40–80% of the critical path delay [35]. In our simulations with a clustered FPGA architecture built using PTM 45nm models [36], we have also seen that net delay contributes on an average 40.3% to the critical delay for MCNC benchmark circuits when mapped to the FPGA framework. The situation is particularly grim since the delay for through the Cu-interconnect is not decreasing, rather increasing with each technology generation. This is in complete contrast with the gate/logic delay which continues to scale with each technology node (refer to Fig. 1.2). The primary reason as explained in [37, 38] is that wire resistance is increasing with each technology node. It is therefore likely that performance for programmable interconnects will continue to remain a serious bottleneck in future FPGA designs [37]. Little can however be done at the circuit level to mitigate the effect of the programmable interconnects. Circuit optimizations in routing switches through buffer and transistor sizing have only demonstrated minimal ($\sim$7%) improvement in delay [39].

Since the interconnect delay is not scaling with technology, the critical path delay for the FPGA platform is not improving appreciably with technology scaling. Our experiments with older and latest generation of Altera FPGAs, as shown in Fig. 2.7 also suggest the poor technological scalability of performance for the conventional FPGA platform. From Fig. 2.7 we note that while the average delay improvement due to technology scaling from 130 to 90nm is 27%, the same from 90 to 65nm is only 13%.

3. **Reliability:** In nanoscale technology, reconfigurable frameworks such as FPGAs have been shown to have increased vulnerability towards logic and interconnect failures. Among the many causes for logic failure, the topic which is highly researched is soft-error in SRAM based FPGA. In scaled technologies, the impact of hit of energized particles is likely to be more severe since it may lead to

multi-bit upsets, thereby causing detection and correction a serious challenge [40]. At circuit level, researchers have proposed different implementations of SRAM [41] which are hardened to soft-error induced failures. Depending on the nature of the soft-error, the effect in the FPGA may be benign or devastating. For example, soft-error induced delay failures can only lead to transient failures. Configuration bits in routing switches however may be inadvertently switched due to soft-error which may even destroy the chip. An effective solution towards soft-error immunity is therefore highly desired. Parametric variations in FPGA interconnects can also lead to delay failures [42] for designs mapped to the FPGA framework. However, no effective circuit techniques have been proposed to mitigate the effect of variability in programmable interconnects across a FPGA device.

2.3.2 Architecture

For reconfigurable frameworks, in general architectural modifications have been found to be more effective than circuit-level optimizations.

1. **Power:** LUT based evaluation of common arithmetic functions such as addition and multiplications were found to be extremely expensive in terms of power and performance. Modern FPGAs therefore include custom adder and multipliers and also memory blocks which significantly improves the area and energy-efficiency for these frameworks [43,44]. Other approaches include FPGA device and architecture co-optimization [45] which explores FPGA architectures with support for V_{dd} programmability.
2. **Performance:** A large number of architectural modifications have been proposed to improve performance in FPGA frameworks. From completely distributed LUT based designs [46], FPGAs have shifted to LUT clusters for improved area and performance efficiency [47]. LUTs of sizes 4–6 and cluster sizes of 4–10 have been shown to be most effective in terms of area-delay product at 0.18 µm [47]. As the technology is scaled further, we believe LUTs with larger input size would be more area efficient. In addition to LUT clusters, modern FPGAs have become heterogenous by incorporating custom adders, multipliers and memory elements. With only LUT based function evaluation, the performance gap between FPGAs and ASICs were estimated to be 40×. This gap however reduces to 21× by incorporating the hard adder and multiplier elements [27]. Finally, to improve the performance of interconnects, researchers have looked into pipelining of FPGA interconnects [48]. Vendors like *Achronix* [49] have demonstrated that asynchronous pipelining of FPGA interconnects can significantly improve the FPGA performance. New FPGA architectures have also been proposed that include segmentation and buffering to optimize speed and density [50].
3. **Reliability:** Architectural support for reliability primarily include automatic detection and correction capability for soft-error induced failures in the FPGA

fabric. These mainly target the configuration bits and the datapath latches in the FPGA framework. Modern FPGAs come with such support [51]. Recent works have also focussed on reducing the soft-error vulnerability of SRAM arrays present on the FPGA [52].

2.3.3 Software

Since applications are mapped off-line onto the FPGA framework, a large number of optimizations are possible in the software for improving the power, performance and reliability of the mapped design.

1. **Power:** A significant amount of works has been done on power-aware application mapping for the FPGA framework. In [53] a novel synthesis approach is proposed for FPGA which improves resource selection and reduces the total number of muxes to effectively reduce FPGA power consumption. In [54], the authors propose a design methodology that implements various computational kernels in an energy efficient manner by characterizing various architectures (e.g. linear array, systolic array) that can implement the same kernel and choosing between the architectures based on performance and resource constraints. Although such an approach can be efficient in reducing the power consumption of applications mapped to platform FPGAs (consisting of hard processors, FPGA fabric, DSP blocks etc.), it does not directly address the power consumption for the reconfigurable framework. Other works target leveraging on the V_{dd} programmability of the FPGA fabric for drastic reduction in the interconnect power [55].
2. **Performance:** With modern FPGAs integrating LUT clusters in their design, application mapping tools have been modified to exploit the spatial locality of the LUTs during application mapping. Moreover, these mapping tools also take advantage of fracturable LUTs for packing of applications [56]. Timing-driven packing, placement and routing algorithms [57] have also been proposed to improve the routing delay for designs mapped to the FPGA framework.
3. **Reliability:** A number of software level optimizations have been proposed to improve the reliability of operation for FPGA frameworks. One of the approaches skews the LUT contents for improving soft-error tolerance in FPGA [41]. Moreover soft-error aware placement [58] has been shown to drastically reduce the number of bits sensitive to soft-error. Furthermore, single event upset aware routing algorithms have also been proposed for reliability improvement in FPGA frameworks [59].

References

1. [Online], "The Landscape of Parallel Computing Research: A View From Berkeley". http://view.eecs.berkeley.edu/wiki/Main_Page
2. F. Pollack, "New Microarchitecture Challenges in the Coming Generations of CMOS Process Technologies", in *Intl. Symp. on Microarchitecture*, 1999
3. E.S. Chung, P.A. Milder, J.C. Hoe, K. Mai, "Single-Chip Heterogeneous Computing: Does the Future Include Custom Logic, FPGAs, and GPGPUs? in *Intl. Symp. on Microarchitecture*, 2010
4. K. Compton, S. Hauck, "Reconfigurable computing: a survey of systems and software". ACM Comput. Surv. **34**(2), 171–210 (2002)
5. R. Hartenstein, "A Decade of Reconfigurable Computing: A Visionary Retrospective", in *DATE*, 2001
6. K. Eguro, S. Hauck, "Resource allocation for coarse-grain FPGA development". IEEE Trans. Comput. Aided Des. Integrat. Circ. Syst. **24**(10), 1572–1581 (2005)
7. V. Betz, J. Rose, A. Marquardt, "Architecture and CAD for Deep-Submicron FPGA". (Springer, Heidelberg, 1999)
8. A. Dehon, "DPGA Utilization and Application", in *Intl. Symp. on FPGAs*, 1996
9. C. Ebeling, D.C. Cronquist, P. Franklin, "RaPiD - Reconfigurable Pipelined Datapath", in *Intl. Workshop on Field-Programmable Logic and Applications*, 1996
10. S. Hauck, T.W. Fry, M.M. Hosler, J.P. Kao, "The Chimaera reconfigurable functional unit". IEEE Trans. Very Large Scale Integrat. Syst. **12**(2), 206–217 (2004)
11. E. Mirsky, A. Dehon, "MATRIX: A Reconfigurable Computing Architecture with Configurable Instruction Distribution and Deployable Resources", in *FPGAs for Custom Computing Machines*, 1996
12. S.C. Goldstein, H. Schmit, M. Moe, M. Budiu, S. Cadambi, R.R. Taylor, R. Laufer, "PipeRench: A Coprocessor for Streaming Multimedia Acceleration", in *Intl. Symp. on Computer Architecture*, 1999
13. C. Brunelli, F. Garzia, J. Nurmi, "A coarse-grain reconfigurable architecture for multimedia applications featuring subword computation capabilities". J. Real Time Image Process. **3**(1), 21–32 (2006)
14. [Online], "The Kress ALU Array". http://xputers.informatik.uni-kl.de/faq-pages/kressalu.html
15. R. Razdan, M.D. Smith, "A High-Performance Microarchitecture with Hardware-Programmable Functional Units", in *Intl. Symp. on Microarchitecture*, 1994
16. T. Miyamori, K. Olukotun, "REMARC: Reconfigurable Multimedia Array Coprocessor", in *Intl. Symp. on FPGAs*, 1998
17. R.D. Wittig, P. Chow, "OneChip: An FPGA Processor with Reconfigurable Logic", in *FPGAs for Custom Computing Machines*, 1996
18. J. Babb et al., "The RAW Benchmark Suite: Computation Structures for General Purpose Computing", in *FPGAs for Custom Computing Machines*, 1997
19. M. Gokhale et al., "SPLASH: A Reconfigurable Linear Logic Array", in *Intl. Conference on Parallel Processing*, 1990
20. J.R. Hauser, J. Wawrzynek, "Garp: a MIPS Processor with a Reconfigurable Coprocessor", in *FPGAs for Custom Computing Machines*, 1997
21. D. Chen, J. Rabaey, "Reconfigurable multi-processor IC for rapid prototyoing of algorithm-specific high-speed datapaths". IEEE J. Solid State Circ. **27**(12), 1895–1904 (1992)
22. [Online], "Tensilica's Xtensa customizable processors". http://www.tensilica.com/products/xtensa-customizable.htm
23. [Online], "Stretch: Software Configurable Processors". http://www.stretchinc.com/
24. H. Singh, M. Lee, G. Lu, F.J. Kurdahi, N. Bagherzadeh, E.M. Chaves Filho, "MorphoSys: an integrated reconfigurable system for data-parallel and computation-intensive applications". IEEE Trans. Comput. **49**(5), 465–481 (2000)
25. F. Barat, R. Lauwereins, G. Deconinck, "Reconfigurable instruction set processors from a hardware/software perspective". IEEE Trans. Softw. Eng. **28**(9), 847–862 (2002)

26. T. Mudge, "Power: a first-class architectural design constraint". IEEE Comput. **34**(4), 52–58 (2001)
27. I. Kuon, J. Rose, "Measuring the gap between FPGAs and ASICs". IEEE Trans. Comput. Aided Des. Integrat. Circ. Syst. **26**(2), 203–215 (2007)
28. L. Shang, A.S. Kaviani, K. Bathala, "Dynamic Power Consumption in Virtex^{TM}-II FPGA Family", in *Intl. Symp. on FPGAs*, 2002
29. T. Tuan, B. Lai, "Leakage Power Analysis of a 90nm FPGA", in *Custom Integrated Circuits Conf.*, 2003
30. F. Li, Y. Lin, L. He, "Field programmability of supply voltages for FPGA power reduction". IEEE Trans. Comput. Aided Des. Intgerat. Circ. Syst. **26**(4), 752–764 (2007)
31. F. Li, Y. Lin, L. He, J. Cong, "Low-Power FPGA Using Pre-defined Dual-Vdd/Dual-Vt Fabrics", in *Intl. Symp. on FPGAs*, 2004
32. J.H. Anderson, F.N. Najm, "Low-Power Programmable Routing Circuitry for FPGAs", in *ICCAD*, 2004
33. A. Gayasen et al., "Reducing Leakage Energy in FPGAs Using Region-Constrained Placement", in *Intl. Symp. on FPGAs*, 2004
34. Y. Lin, F. Li, L. He, "Routing Track Duplication with Fine-Grained Power-Gating for FPGA Interconnect Power Reduction", in *ASP-DAC*, 2005
35. S. Das, A.P. Chandrakasan, A. Rahman, R. Reif, "Wiring requirement and three-dimensional integration technology for field programmable gate arrays". IEEE Trans. Very Large Scale Integrat. Syst., **11**(1), 44–54 (2003)
36. [Online], "Predictive Technology Model". http://ptm.asu.edu/
37. [Online], "Improving FPGA Performance and Area Using an Adaptive Logic Module". www.altera.com/literature/cp/cp-01004.pdf
38. [Online], "ITRS 2007 : Interconnect". http://www.itrs.net/links/2007itrs/home2007.htm
39. G. Lemieux, D. Lewis, "Circuit Design of Routing Switches", in *Intl. Symp. on FPGAs*, 2002
40. S. Paul, F. Cai, X. Zhang, S. Bhunia, "Reliability-driven ECC allocation for multiple bit error resilience in processor cache". IEEE Trans. Comput. **60**(1), 20–34 (2011)
41. S. Srinivasan, A. Gayasen, N. Vijaykrishnan, M. Kandemir, Y. Xie, M.J. Irwin, "Improving Soft-Error Tolerance of FPGA Configuration Bits", in *ICCAD*, 2004
42. Y. Lin, L. He, M. Hutton, "Stochastic physical synthesis considering prerouting interconnect uncertainty and process variation for FPGAs". IEEE Trans. Very Large Scale Integrat. Syst. **16**(2), 80–88 (2008)
43. [Online], "FPGAs Provide Reconfigurable DSP Solutions". www.altera.com/literature/wp/wp_dsp_fpga.pdf
44. [Online], "DSP Co-Processing in FPGAs: Embedding High-Performance, Low-Cost DSP Functions". www.xilinx.com/support/documentation/white_papers/wp212.pdf
45. L. Cheng, P. Wong, F. Li, Y. Lin, L. He, "Device and Architecture Cooptimization for FPGA Power Reduction", in *DAC*, 2005
46. J. Rose, R.J. Francis, D. Lewis, P. Chow, "Architecture of field programmable gate arrays: The effect of logic functionality on area efficiency". IEEE J. Solid State Circ. **25**(5), 1217–1225 (1990)
47. E. Ahmed, J. Rose, "The effect of LUT and cluster size on deep submicron FPGA performance and density". IEEE Trans. Very Large Scale Integrat. Syst. **12**(3), 288–298 (2004)
48. A. Sharma, K. Compton, C. Ebeling, S. Hauck, "Exploration of Pipelined FPGA Interconnect Structures", in *Intl. Symp. on FPGAs*, 2004
49. [Online], "Achronix Semiconductor Corp." http://www.achronix.com/
50. V. Betz, J. Rose, "FPGA Routing Architecture: Segmentation and Buffering to Optimize Speed and Density", in *Intl. Symp. on FPGAs*, 1999
51. [Online], "Error Detection and Recovery Using CRC in Altera FPGA Devices". http://www.altera.com/literature/an/an357.pdf
52. S.P. Park, D. Lee, K. Roy, "Soft-error-resilient FPGAs using Built-2-D hamming product code". IEEE Trans. Very Large Scale Integrat. Syst., 248–256 (2011)

53. D. Chen, J. Cong, Y. Fan, "Low-Power High-Level Synthesis for FPGA Architectures", in *ISLPED*, 2003
54. V.K. Prasanna, "Energy-efficient computations on FPGAs". J. Supercomput. **32**(2), 139–162 (2005)
55. Y. Hu, Y. Lin, L. He, T. Tuan, "Physical synthesis for FPGA interconnect power reduction by dual-vdd budgeting and retiming". ACM Trans. Des. Autom. Electron. Syst. **13**(2), 1–29 (2008)
56. [Online], "Stratix IV FPGA ALM Logic Structure's 8-Input Fracturable LUT". http://www.altera.com/products/devices/stratix-fpgas/stratix-iv/overview/architecture/stxiv-alm-logic-structure.html
57. [Online], "VPR and T-VPack 5.0.2 Full CAD Flow for Heterogeneous FPGAs". http://www.eecg.utoronto.ca/vpr/
58. M.A. Abdul-Aziz, M.B. Tahoori, "Soft Error Reliability Aware Placement and Routing for FPGAs", in *ITC*, 2010
59. S. Golshan, E. Bozorgzadeh, "Single-Event-Upset (SEU) Awareness in FPGA Routing", in *DAC*, 2007

Chapter 3
Motivation for a Memory-Based Computing Hardware

Abstract In this chapter we first provide a summary of the desired characteristics that a compute framework must have to overcome the challenges faced by conventional hardware and software reconfigurable frameworks at nanoscale technologies. We then provide an outline for a new computing model which bridges the gap between memory and logic. Given the rapid evolution of CMOS and non-CMOS memory technologies, we explain the benefits of such in-memory computing model.

3.1 Motivation for a New Computing Model

The technology trends as put forward in the previous chapters point to following simple conclusions—(i) *Performance can be achieved through parallelism*. Although Amdahl's law puts an upper bound on the maximum speed-up that can be achieved for an application through parallel-operation, J. Gustafson correctly noted that for massively parallel machines, it is possible to compute large data sets in the same amount of time [1]. (ii) *Efficiency is in locality*. Clearly the performance and energy overhead due to movement of data can easily surpass the energy spent in useful computations. Hence only localized computation with minimum data transport can keep the energy efficiency improving with technology scaling. This only can be achieved by co-designing the hardware and the software [2]. It is therefore worthwhile to explore the design of novel computing frameworks which:

- improve the energy efficiency for current reconfigurable frameworks by reducing the contribution from the programmable interconnects.
- leverage on the benefits offered by emerging non-CMOS memory technologies.
- are robust to failures resulting from parametric variation.

The above requirements point to a resilient framework which can be as energy-efficient as a hardware reconfigurable framework, but scales as well as a software reconfigurable one. Moreover, the new framework must address the off-chip performance and energy bottleneck by computing in close proximity to off-chip non-volatile storage. Following are two key motivations for such a in-memory computing model.

3.1.1 Hardware Acceleration

The above requirements apply to any general purpose computing framework. With growing demand for low-power high-performance application execution, these *hardware accelerators* are playing a major role in speeding up diverse applications from different application domains [4–7]. FPGAs, graphical processing units (GPUs) are hardware acceleration devices which are commonly used today. A study of the algorithm *hotspots* which are commonly mapped to these hardware accelerators is presented in [8]. These include dense and sparse matrix operations, signal transforms, mapreduce, graph traversal, dynamic programming, structured and unstructured grid etc. A look at the performance limitation of these *dwarfs* on software reconfigurable system reveals that most of these are limited by memory bandwidth and latency, rather than by computing resources. Common *hardware accelerators* such as FPGAs and GPUs partly solve this problem by integrating an on-chip memory hierarchy and having many operations in parallel to effectively hide the data transfer delay.

3.1.2 Opportunities Provided by Emerging Technologies

In the quest of a potential alternative to CMOS at the end of its roadmap [9], multitude of research efforts have been directed towards investigating novel devices with interesting and unique switching characteristics. These emerging research devices are primarily classified into two types:

- *Memory devices:* Some of the promising memory devices include (a) spin torque transfer random access memory (STTRAM) [10], (b) phase change random access memory (PCRAM) [11], (c) resistive random-access memory (RRAM) [12] (d) nanoelectromechanical memory (NEMM) [13] and (e) molecular memory [14]. One important characteristic that distinguishes the emerging memory devices from existing memory technologies such as static random access memory (SRAM) and dynamic random access memory (DRAM) are that all the above memories are non-volatile. This implies that configuration can be stored indefinitely in these memories without requiring any supply, consequently facilitating low-power applications.
- *Logic and Alternative Information Processing Devices:* These array of devices include (a) single-electron transistors (SET) [15], (b) carbon nanotube field effect transistor (CNTFET) [15], (c) semiconductor nano-wires, (d) quantum-dot cellular automata (QCA) [15] and (e) molecular transistors [15].

3.1 Motivation for a New Computing Model

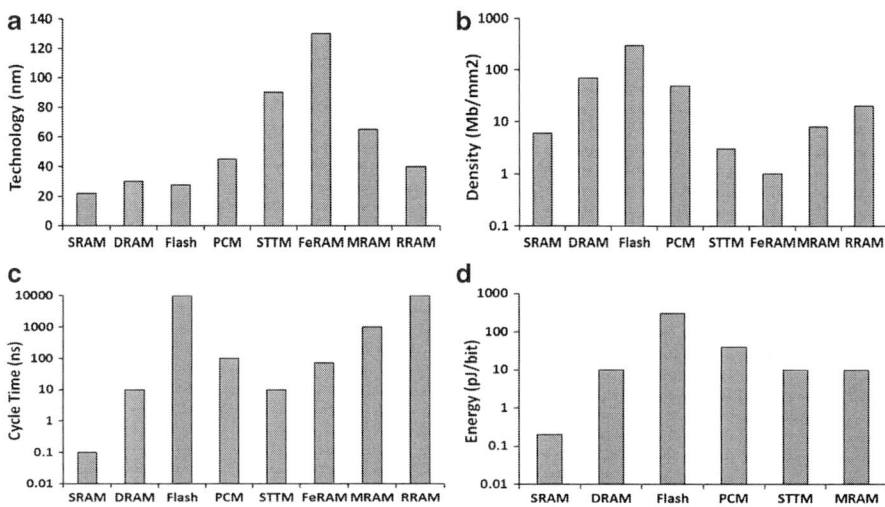

Fig. 3.1 Comparison of current and emerging memory technologies [3]

Although most of these emerging nanodevices are still in their infancy, they hold tremendous potential in terms of integration density (in the order of 10^{10} devices/cm^2) [15]. low power operation and higher switching speed. An extensive review of these emerging devices can be obtained in [15]. In spite of differences in the working of these individual devices, they bear the following common characteristics:

- Each of these technologies are still at their infancy and require interfacing with conventional CMOS technology for reliable operation.
- Inexpensive fabrication methodology which can support large scale manufacturing of nanoscale devices with reproducible behavior is still being investigated.
- Both the emerging memory and logic devices are characterized by high density and are more amenable towards the fabrication of regular structures.
- Due to limitations in fabrication techniques, these nanoscale devices are characterized by high defect rates and large variability.

Figure 3.1 compares the conventional volatile and non-volatile memory technologies with the emerging non-volatile memory technologies which have demonstrated considerable promise [3]. Technologies such as STTM has been shown to have better performance and energy, as well as endurance compared to conventional non-volatile Flash memory technology. As these technologies improve, they can potentially replace one or more volatile or non-volatile memory technologies which have been in commercialization for decades. In lieu of the above trends, it is necessary that a computing model beneficial for the emerging nanodevices should bear the following attributes:

- In order to leverage on the advantages such as density and low power of operation as offered by the nanodevices, it is desired that computation is actually performed using these nanodevices and CMOS is used for purposes such as level restoration and interconnection between computing blocks.
- Interface between CMOS and the nanoscale devices should be few and well defined. Due to inherent difficulty in alignment and accurate patterning of the emerging nanodevices, the computing model must be able tolerate any variability in the CMOS-nano interface.
- The model should also account for the large defect rate for the emerging nanodevices and provide mechanisms of defect tolerance without any significant overhead.

In perspective of the attributes required for the nanocomputing models, memory based reconfigurable computing has emerged as one of the primary choice. The primary advantages offered by memory as a computing platform in nanoscale architectures are:

- As mentioned emerging nanodevices are highly amenable for dense and regular circuit design. Recently, researchers have reported successful fabrication of 160 Kbit of molecular memory patterned at 10^{11} bits per square centimeter [16]. Such dense memory arrays are realized using either by self-assembly process or by nano-imprinting method [15].
- Memory preserves the density advantage offered by the emerging nanodevices. Since these nanodevices need to be interfaced with CMOS logic, a dense nanoscale memory structure can substantially reduce the CMOS overhead for a single memory bit. Moreover, memory structures realized using nanodevices have very well defined CMOS interfaces thus facilitating CMOS-nano hybridization.
- It is well known that redundancy or error-correction based defect tolerance schemes for random logic incur significant overhead. However, due to the regular nature of the memory, high defect rates can be tolerated by either re-mapping to non-defective locations, by the use of redundant rows and columns or by using error correction schemes [15]. Overhead incurred per bit is much less in the latter case than for random logic.

3.2 Proposed Solution and Benefits

In this work, we propose a novel memory based hardware reconfigurable framework (MBC framework) to address the challenges of technological scaling as summarized in the previous sections. The model is independent of the underlying memory technology as long it satisfies memory read and write operations. The memory based computing model thus applies to both on-chip as well as off-chip memory. In particular, the framework has the following key advantages:

- The framework is spatio-temporal in nature. This increases the percentage of local computation at each compute node of the proposed framework, thus minimizing the data movement across the framework and improving the energy-efficiency. With effective contribution from programmable interconnects reduced, this framework enjoys a better technological scalability compared to fully spatial reconfigurable frameworks.
- The basic computing fabric for the framework is memory. Computations of arbitrary complexity can be suitably partitioned and mapped as Look-up Tables (LUTs) to this memory array which also stores the data which is to be processed. Computation and storage are therefore at the same location which mitigates the well-known Von-Neumann bottleneck.
- Since the main computing fabric is memory, the framework can leverage on the benefits of a number of emerging non-volatile memory technologies, namely large integration density, low power, high read speed and non-volatility.
- Programming the proposed framework is facilitated through the use of standard instructions and the input application is automatically scheduled by the software flow in space and time to maximize the energy benefit of temporal computing without sacrificing the performance benefit of spatial computing. Data movement across the storage hierarchy is also explicitly controlled by the application mapping tool.
- Leveraging on the fact that memory arrays can be easily reconfigured around defects and failures by using redundant rows and columns [17], the proposed framework improves reliability of operation in nanoscale technologies.
- The proposed framework can act as a hardware accelerator for compute-intensive applications by mapping complex datapath as well as to the data-intensive applications by storing the data to be processed inside the same framework and leveraging on the data high-bandwidth available inside memory.

As opposed to a pure software based execution which is often non-optimal for many algorithmic tasks, the proposed hardware reconfigurable framework can improve performance and energy through techniques such as pipelining and parallel processing. As opposed to conventional FPGA frameworks which are fully spatial and are dominated by programmable interconnect, the proposed framework is spatio-temporal and reduces the contribution from interconnects. Finally unlike the GPU, it can map arbitrary applications using Look-up tables and perform bit-level computations which imparts higher flexibility compared to a GPU platform.

References

1. J.L. Gustafson, "Reevaluating Amdahl's Law". ACM Comm. **31**(5), 532–533 (1988)
2. [Online], "Future processing : Extreme scale". http://www.hpcuserforum.com/EU/downloads/DARPAExtremscaleHarrod_ACS_Bri%ef_01709.pdf
3. "Assessment of the Potential & Maturity of Selected Emerging Research Memory Technologies" http://www.itrs.net/links/2010itrs/2010Update/ToPost/ERD_ERM_2010FINALReportMemoryAssessment_ITRS.pdf

4. H. Singh, M. Lee, G. Lu, F.J. Kurdahi, N. Bagherzadeh, E.M. Chaves Filho, "MorphoSys: an integrated reconfigurable system for data-parallel and computation-intensive applications". IEEE Trans. Comput. **49**(5), 465–481 (2000)
5. S.C. Goldstein, H. Schmit, M. Moe, M. Budiu, S. Cadambi, R.R. Taylor, R. Laufer, "PipeRench: A Coprocessor for Streaming Multimedia Acceleration", in *Intl. Symp. on Computer Architecture*, 1999
6. S. Yehia, N. Clark, S.A. Mahlke, K. Flautner, "Exploring the Design Space of LUTbased Transparent Accelerators", in *CASES*, 2005
7. A. Agarwal et al., "A 320mV-to-1.2V On-Die Fine-Grained Reconfigurable Fabric for DSP/Media Accelerators in 32nm CMOS", in *Intl. Solid-State Circuits Conference*, 2010
8. [Online], "The Landscape of Parallel Computing Research: A View From Berkeley". http://view.eecs.berkeley.edu/wiki/Main_Page
9. [Online], "ITRS 2007 : Interconnect". http://www.itrs.net/links/2007itrs/home2007.htm
10. T.W. Andre et al., "A 4-Mb 0.18-m 1T1MTJ Toggle MRAM with Balanced Three Input Sensing Scheme and Locally Mirrored Unidirectional Write Drivers", in *Intl. Solid-State Circuits Conference*, 2004
11. W.Y. Cho et al., "A 0.18-m 3.0-V 64-Mb nonvolatile phase-transition random access memory (PRAM)". IEEE J. Solid State Circ. **40**, 293–300 (2005)
12. S.T. Hsu, T. Li, N. Awaya, "Resistance random access memory switching mechanism". Appl. Phys., 024517–024517-8 (2007)
13. T. Rueckes et al., "Carbon nanotube-based nonvolatile random access memory for molecular computing" Science. **289**(5476), 94–97 (2000)
14. W. Wu et al., "One-kilobit cross-bar molecular memory circuits at 30-nm half-pitch fabricated by nanoimprint lithography". Appl. Phys., 1173–1178 (2005)
15. S. Paul, S. Bhunia, "Computing with Nanoscale Memory: Model and Architecture", in *Intl. Symp on Nanoscale Architecture*, 2011
16. M.R. Stan et al., "Molecular electronics: from devices and interconnect to circuits and architecture". Proc. IEEE **91**, 1947–1957 (2003)
17. S. Mukhopadhyay, H. Mahmoodi, K. Roy, "Modeling of failure probability and statistical design of SRAM array for yield enhancement in nanoscaled CMOS". IEEE Trans. Comput. Aided Des. Integrat. Circ. Syst. **24**(12), 1859–1880 (2005)

Part II
Memory Based Computing: Overview

Chapter 4
Key Features of Memory-Based Computing

Abstract In this chapter, we highlight the key features of the memory-based computing (MBC) framework. We differentiate the proposed computing model from previously proposed in-memory computing architectures. We describe the application scenarios for which MBC is likely to offer benefit over conventional hardware and software programmable frameworks. Lastly we summarize the key features of the proposed MBC model.

4.1 Computing in Memory Versus Computing with Memory

Computing in memory or processing in memory was made popular by projects [1–3] which attempted to mitigate the Von-Neumann bottleneck. In all these approaches, a software reconfigurable framework was placed in close proximity with DRAM based off-chip main memory so that the performance and energy overhead in transferring the data and the instruction from the off-chip memory to the on-chip memory is minimized. The large data bandwidth available in off-chip memories such as DRAM is exploited by instantiating vector processors which can perform vector operations over the data which is loaded in parallel from the memory. Efforts have also been made in integrating DRAM memory with reconfigurable logic [3] so that hardware reconfigurable systems can also leverage on the concept of *Computing in memory*. Note that however, in all these systems, the memory is only used to hold data and in some cases instructions.

Computing with memory is a completely different concept from *Computing in memory* since it actually uses the memory arrays for computing. In order to explain the use of memory in computation, we will try to answer the following questions:

1. **How can we perform computation in memory?**

Ans. The main idea is to partition to input application into smaller control and data flow graphs, map the individual graphs using Look up Tables (LUTs) and map these LUTs to the memory array. The memory then holds not only the data, but also the function which will operate on the data. Note that a single memory array can hold one LUT, in which case the MBC framework is fully spatial or multiple LUTs can be mapped to a single memory array and evaluated over multiple clock cycles, thus creating a temporal computing model.

2. Which functions should we compute in memory?

Ans. Since the size of a LUT grows exponentially with the number of inputs and linearly with the number of outputs, it is desired that the function which is mapped as a LUT to the memory array has relatively small number of inputs and outputs. If the number of inputs and outputs to the original function is very large, then it is compulsory that the function may be bit-sliced into smaller operations. Arithmetic operations such as addition and multiplication and logic operations such as "and", "xor" etc. can be easily bitsliced and mapped as LUTs to the memory array. The caveat over here is however, delay and energy for executing these arithmetic and logic operations in a custom logic is much smaller than compared to their evaluation using a LUT based approach with the LUT mapped to the memory array. However, the effectiveness of the memory based computing approach becomes more prominent for complex functions such as trigonometric and transcendental functions [4]. It has been demonstrated that it is possible to decompose these functions in a spatial manner and evaluate them using multiple LUTs [5]. The area and delay advantage of memory based computing has also been shown to hold true for functions which comprise of random logic. In case of mapping such functions to a LUT based hardware reconfigurable framework such as FPGA, it is possible to *fuse* multiple small logic functions and form an arbitrary complex function [6, 7] which is mapped to the memory array as a multi-input multi-output LUT. This minimizes the total number of operations in the function, thereby improving the area and performance for the mapped application. *In summary, applications which have complex datapath or comprises of simple operations which can be merged into a complex function with limited number of inputs and outputs are amenable to memory based mapping.*

3. When do we perform memory based computation?

Ans. The idea of memory based computing applies to both software and hardware reconfigurable systems. For software reconfigurable systems, the idea of storing the LUTs in memory for function evaluation have been applied in diverse contexts such as: (i) *Performance Improvement:* LUTs stored in on-chip memory have been widely used for performance improvement in software reconfigurable systems. Their primary use is in storing the pre-calculated results for: (i) complex trigonometric functions; (ii) transformed image pixel in image processing applications such as RGB to grayscale conversion and (iii) complex logic function such as *population function* [8]. Computing the results for such functions in software is expensive in terms of performance. Looking up the pre-computed value and possibly later interpolating two or more values to achiever higher accuracy is much more beneficial is such scenarios. (ii) *Reliability Improvement:* In [9], we have demonstrated that it is possible to use memory based computing for improving the reliability for modern processor systems. The idea is to bitslice common arithmetic operations such as addition and multiplication and store the corresponding LUTs in on-chip cache. In case, the custom adder and multiplier inside the processor is faulty (either due to manufacturing defects or parametric variation), these operations can be migrated to the on-chip memory and calculated over multiple clock cycles. We however noted that the abundance of narrow width operands in general purpose applications reduces the performance overhead for memory based computation and improves the cycles per instruction (CPI) for a processor system with faulty execution units.

Memory based computing is more effective for the case of hardware reconfigurable frameworks which often partition the input application into smaller partitions and represent these individual partitions using LUTs. For both spatial [6, 7] and spatio-temporal hardware reconfigurable frameworks [10–12], memory based computing have been shown to improve the area, delay and energy for the mapped application compared to mapping the input application to small 1-D distributed LUTs. The unique feature of a memory based hardware reconfigurable framework is that in such a framework, fabrics for computation and data storage are same, i.e. *memory*. It is therefore easier to realize a *Computing with memory* hardware framework on a *Computing in memory* platform. The rewards of such an integration is significant since the delay and energy associated with movement of data into and out from the memory is completely eliminated.

4.2 Highlights of Memory-Based Computing Framework

The key highlights of the memory-based computing hardware are:

- In contrast to large on-chip caches, each processing element of the framework incorporates a small local memory which scales better with technology. The local memory not only stores data but also multi-input multi-output LUT responses for functions which are too complex to realize in logic. This computing with memory [12–14] approach has been demonstrated to be extremely attractive for emerging nanoscale non-CMOS memories (such as STTRAM, molecular memory) which are particularly amenable for dense non-volatile memory design.
- Significant gains in energy-efficiency can be obtained by computing inside the NVM, instead of transporting the data stored in NVM to an external compute hardware. MBC is an attractive low-overhead and energy-efficient candidate for such in-memory computing. In the NVM based MBC model, multiple NVM arrays (e.g. Flash blocks) can be grouped together to form a single processing element. Each PE would process its local data, communicating with other PEs as per application requirement.
- MBC is a spatio-temporal framework. In contrast to conventional FPGAs which are fully spatial and are limited by interconnect power and performance, MBC allows the application mapping tool to trade-off distributed vs local execution for maximum energy-efficiency.

- MBC is a mixed-granular framework capable of exploiting both data and task parallelism. The architecture abstracts the low-level hardware implementation details from the programmer. The application mapping tool can receive a high-level application description and efficiently map it to an underlying hardware framework. In this respect, it is easily programmable like a GPU and flexible like a FPGA.

References

1. P.M. Kogge, T. Sunaga, H. Miyataka, K. Kitamura, E. Retter, "Combined DRAM and Logic Chip for Massively Parallel Systems", in *Conf. on Advanced Research VLSI*, 1995
2. [Online], "Computational RAM". http://www.eecg.toronto.edu/~dunc/cram/
3. [Online], "The Berkeley Intelligent RAM (IRAM) Project". http://iram.cs.berkeley.edu/
4. A. Dehon, "Reconfigurable Architectures for General-Purpose Computing". Technical report, MIT, 1996
5. T. Sasao, S. Nagayama, J.T. Butler, "Numerical function generators using LUT cascades". IEEE Trans. Comput. **56**(6), 826–838 (2007)
6. J. Cong, S. Xu, "Technology Mapping for FPGAs with Embedded Memory Blocks", in *Intl. Symp. on FPGAs*, 1998
7. S.J.E. Wilton, "SMAP: Heterogeneous Technology Mapping for Area Reduction in FPGAs with Embedded Memory Arrays", in *Intl. Symp. on FPGAs*, 1998
8. [Online], "Lookup table". http://en.wikipedia.org/wiki/Lookup_table
9. S. Paul, S. Bhunia, "Dynamic transfer of computation to processor cache for yield and reliability improvement". IEEE Trans. Very Large Scale Integrat. Syst., 1368–1379 (2011)
10. S. Paul, S. Chatterjee, S. Mukhopadhyay, S. Bhunia, "A Circuit-Software Co-design Approach for Improving EDP in Reconfigurable Frameworks", in *ICCAD*, 2009
11. S. Paul, S. Chatterjee, S. Mukhopadhyay, S. Bhunia, "Nanoscale Reconfigurable Computing Using Non-Volatile 2-D STTRAM Array", in *Intl. Conf. on Nanotechnology*, 2009
12. S. Paul, S. Bhunia, "Memory Based Computing: Reshaping the Fine-Grained Logic in a Reconfigurable Framework", in *Intl. Symposium on Field Programmable Gate Arrays*, 2011
13. S. Paul, S. Bhunia, "A scalable memory-based reconfigurable computing framework for nanoscale crossbar". IEEE Trans. Nanotechnology **9**(2), 451–462 (2010)
14. S. Paul, S. Chatterjee, S. Mukhopadhyay, S. Bhunia, "Energy-efficient reconfigurable computing using a circuit-architecture-software co-design approach". IEEE J. Emerg. Sel. Top. Circ. Syst. (JETCAS) Special Issue Adv. Des. Energy Efficient Circ. Syst., 369–380 (2011)

Chapter 5
Overview of Hardware and Software Architectures

Abstract Memories are typically associated with data storage in computer system. They either store data temporarily (for example by volatile memories such as SRAM, DRAM etc.) or permanently (for example by non-volatile memories such as Flash, magnetic disks etc.). In this chapter we first draw the outline how the embedded memory may be used for computing as well in a time-multiplexed hardware reconfigurable system. In particular, this chapter makes the following key contributions:

- We propose a hardware architecture to utilize a dense 2-D memory array for the purpose of reconfigurable computing. The main idea is to map multiple multi-input multi-output LUTs to the embedded memory array at each compute block and evaluate them over multiple clock cycles. The proposed framework is therefore a spatio-temporal framework unlike the conventional hardware reconfigurable frameworks which are fully spatial. In the proposed hardware architecture, multiple computing elements communicate with each other over a time-multiplexed programmable interconnect.
- Along with the hardware architecture we outline an effective software framework which maps an input application to the proposed hardware reconfigurable framework. The software flow outlined partitions the input application and maps it to multiple compute blocks, and finally places and routes the mapped design.
- We describe the application domains of the proposed memory based computing model. We identify that the model can be applied to realize a stand-alone reconfigurable framework for mapping random logic as well as it can be used for hardware acceleration of algorithmic tasks.

5.1 Overview of Hardware Organization and Software Flow

In light of the above benefits, we have proposed a novel memory based computing (MBC) model. The hardware reconfigurable framework which embodies the proposed model is hereafter referred to as the memory based computing framework.

The key highlights of the proposed computing model are as follows:

1. It maps the input application as multiple multi-input multi-output LUTs and stores these LUTs inside a dense 2-D memory array. The application is then executed by accessing these LUTs topologically over multiple clock cycles. In doing so it mimics a temporal computing framework.
2. For sufficiently large applications which cannot be mapped to a single 2-D memory array, the model proposes a methodology to suitably partition the input application and map them spatially across compute elements, each of which is temporal in nature.

Figure 5.1 illustrates the memory based computing model. The target application is partitioned into multi-input multi-output LUTs which are mapped to dense memory arrays (referred to as as function table) inside each MLB. Information regarding the address, scheduling steps and connectivity among the partitions is stored in a microcode format in a small ip op array (referred to as schedule table) during the application mapping phase. A register file holds the intermediate partition outputs. The schedule table, the function table and the register file forms the core of the MLB. Inside each MLB, the partitions are evaluated in a topological manner over multiple cycles. The requirement for communication between multiple MLBs arises when dependent partitions are mapped to two different MLBs. Inter-MLB communication is achieved through a FPGA-like PI framework. An automated software flow supports the mapping of the CDFG of the target application to the MBC framework. The software flow is responsible for partitioning the CDFG into subgraphs, scheduling them based on available resources and distributing the subgraphs among multiple MLBs. It also performs the complex job of placing the MLBs and routing signals across these MLBs. Upon successful placement and routing, it generates the bitstream required to configure individual MLBs and PI.

5.2 Distinction with Existing Hardware Frameworks

5.2.1 Comparison with Conventional FPGAs

MBC bears marked distinction from conventional FPGA frameworks from the fact that the latter is a fully spatial computing framework which maps the input application as distributed small 1-D LUTs. As a consequence, FPGAs require an elaborate PI network for placement and routing of the LUTs. Although the requirement for interconnect has been reduced by replacing the individual LUTs with LUT clusters, the programmable interconnect has still been observed to contribute to 80% of the power, 60% of the delay and the 75% of the FPGA area in nanoscale technologies [2]. The proposed MBC model differs from the conventional FPGAs on the following points:

5.2 Distinction with Existing Hardware Frameworks

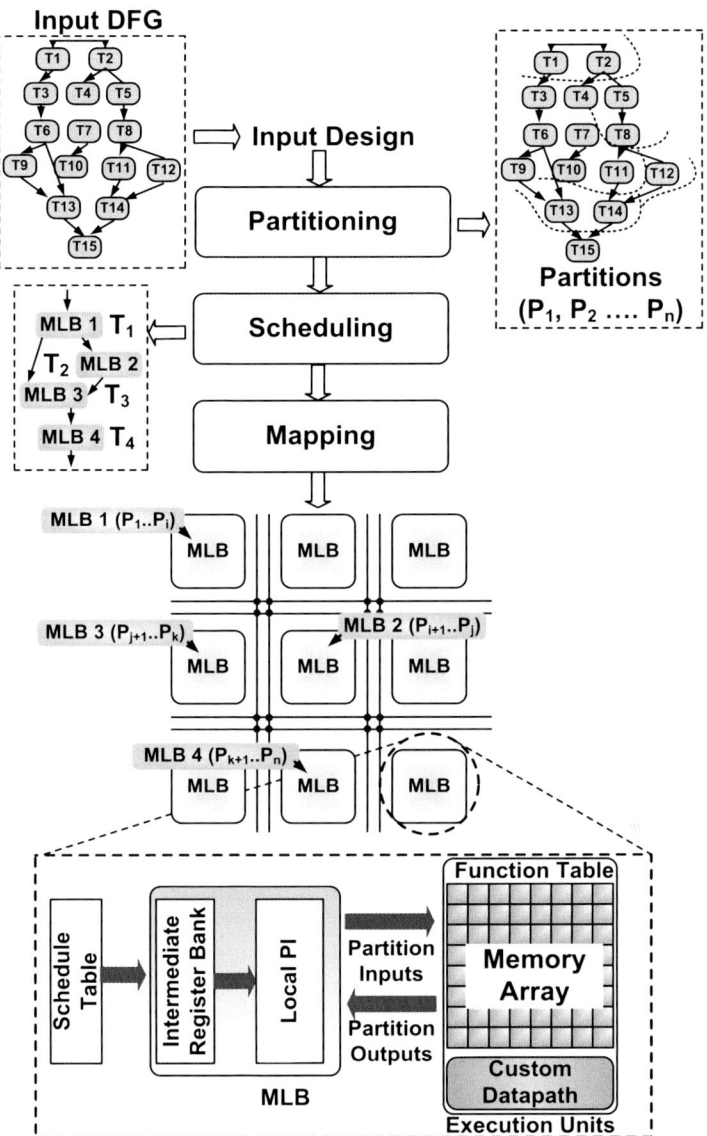

Fig. 5.1 Functional block diagram of memory based computing

1. While FPGA is a fully spatial computing model, MBC is a spatio-temporal model which minimizes the requirement for programmable interconnect by multi-cycle localized execution inside each compute element of the framework.
2. In FPGAs, the input application is mapped to small 1-D LUTs, while the same is mapped as multi-input multi-output LUTs stored in dense 2-D memory arrays in the MBC framework.

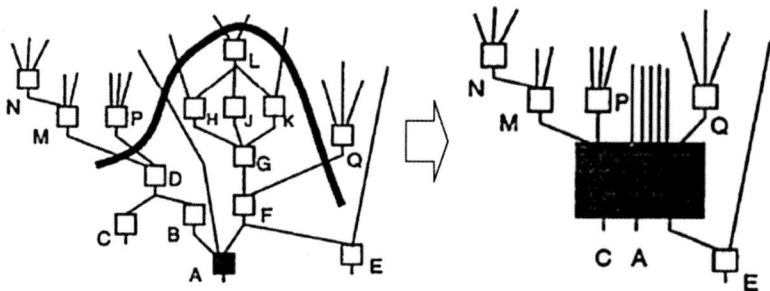

Fig. 5.2 Packing of logic nodes into multi-input multi-output partitions mapped to embedded memory blocks [1]

3. Multi-MLB communication in the MBC framework utilizes the programmable interconnect in a time-multiplexed manner. FPGAs being fully spatial cannot leverage on the benefits of time-multiplexing the PI.

5.2.2 Comparison with EMB Based FPGAs

As emphasized in [1,3,4], the embedded memory arrays inside conventional FPGA fabrics can be used for logic computations when they are not configured as on-chip memories. Figure 5.2 shows the packing of multiple smaller LUTs to larger multi-input multioutput LUTs mapped to the memory array. Such a mapping of smaller LUTs to larger embedded memory arrays can improve (i) the total area required to map a given netlist [1, 3]; (ii) a delay-oriented mapping algorithm [4] achieves significant improvement in performance of the mapped application. We refer to these frameworks as "EMB based Heterogenous FPGA". Although these frameworks employ conventional two-dimensional (2-D) memory array for logic computation, it relies on a purely spatial computing model similar to that of conventional FPGA. The proposed MBC framework is similar but still different from EMB based heterogenous FPGAs on the following points:

1. Similar to EMB based heterogenous FPGAs, MBC also partitions the input application as multi-input multi-output LUTs and maps them to embedded 2-D memory arrays.
2. Due to the use of larger multi-input multi-output LUTs, EMB based heterogenous FPGAs have less interconnect overhead as compared to conventional FPGAs. However, it is still a fully spatial framework and therefore cannot leverage on the time-multiplexing of the PI as can be achieved in the spatio-temporal MBC framework.
3. In EMB based heterogenous FPGAs, a single LUT is mapped to a 2-D memory array. To leverage on the integration densities of nanoscale memory technologies,

current memory arrays are typically large (8–32KB) which allows multiple multi-input multi-output LUTs to the same 2-D memory array. A EMB based heterogenous FPGA cannot therefore benefit from such a scenario. A temporal model at each MLB of the MBC framework allows multiple LUTs to be stored inside the memory array and evaluated in a topological manner over multiple clock cycles.

5.2.3 Comparison with Time-Multiplexed Hardware Reconfigurable Frameworks

A number of time-multiplexed hardware reconfigurable frameworks have been proposed till date [5–7]. The main idea in all these frameworks is to store the configuration for multiple applications in a small memory local to each compute element and routing switch so that it is possible to migrate from one application execution to another with minimum reconfiguration overhead. As demonstrated in [7], such a framework can significantly improve logic utilization on a hardware reconfigurable framework through temporal pipelining and even improve performance. A number of partitioning [8] and scheduling algorithms [5] have also been proposed for such time-multiplexed reconfigurable frameworks. MBC bears some similarities and certain differences with these time-multiplexed frameworks. They are as follows:

1. Similar to the time-multiplexed architecture, configuration bits for a number of applications can be loaded into the each compute block MBC framework and they may be executed in a time-multiplexed fashion.
2. Unlike time-multiplexed frameworks, MBC however does not allow time-multiplexed inter-MLB routing switches. Hence only those applications which have similar interconnection graph can be mapped to the MBC framework and executed in a time-multiplexed fashion.
3. For a single application, previous time-multiplexed frameworks would still behave as a fully spatial framework and therefore incur large overhead in programmable interconnect.
4. In [5–7], the input application is still mapped to small 1-D LUTs at each compute node. On the other hand MBC advocates a temporal execution model at each computing node by mapping multiple LUTs to a dense 2-D memory array.

5.2.4 Comparison with Spatio-Temporal Hardware Reconfigurable Frameworks

A number of spatio-temporal hardware reconfigurable frameworks have been proposed till date [9–11]. All these however are coarse grained reconfigurable frameworks which perform multi-cycle execution at each compute node by incorporating wide arithmetic logic unit (ALU) and storing the configuration bits in a small local memory at each compute node. Such frameworks also incorporate scheduled routing switches [12] similar to time-multiplexed reconfigurable fabrics. Note that communication in such frameworks is often on a word-basis (32-bits), so that the overhead for scheduling the interconnect is often amortized over the buswidth. MBC has the following differences with the spatio-temporal frameworks proposed till date:

1. As opposed to [9–11], MBC is a fine-grained reconfigurable framework. In contrast to [9–11] which incorporate an ALU for multi-cycle execution, in the MBC framework, multi-input multi-output LUTs representing the input application are mapped to a dense 2-D memory array for the purpose of temporal computing.
2. Interconnect is statically scheduled in case of MBC. Hence, MBC poses unique placement and routing challenges, different from those addressed in [12].

References

1. S.J.E. Wilton, "SMAP: Heterogeneous Technology Mapping for Area Reduction in FPGAs with Embedded Memory Arrays", in *Intl. Symp. on FPGAs*, 1998
2. S. Das, A.P. Chandrakasan, A. Rahman, R. Reif, "Wiring requirement and three-dimensional integration technology for field programmable gate arrays". IEEE Trans. Very Large Scale Integrat. Syst. **11**(1), (2003)
3. J. Cong, S. Xu, "Technology Mapping for FPGAs with Embedded Memory Blocks", in *Intl. Symp. on FPGAs*, 1998
4. J. Cong, S. Xu, "Performance-driven technology mapping for heterogeneous FPGAs". IEEE Trans. Comput. Aided Des. Integrat. Circ. Syst. **19**(11), 1268–1281 (2000)
5. S. Trimberger, "Scheduling Designs into a Time-Multiplexed FPGA", in *Intl. Symp. on FPGAs*, 1998
6. D. Jones, D.M. Lewis, "Time-Multiplexed FPGA Architecture for Logic Evaluation", in *Custom Integrated Circuits Conference*, 1995
7. A. Dehon, "DPGA Utilization and Application", in *Intl. Symp. on FPGAs*, 1996
8. G. Wu, J. Land, Y. Chang, "Generic ILP-based approaches for time-multiplexed FPGA partitioning". IEEE Trans. Comput. Aided Des. Integrat. Circ. Syst. **20**(10), 1266–1274 (2001)
9. [Online], "Mosaic Developing power-efficient coarse-grained reconfigurable architectures and tools". http://www.cs.washington.edu/research/projects/lis/www/mosaic/
10. H. Singh, M. Lee, G. Lu, F.J. Kurdahi, N. Bagherzadeh, E.M. Chaves Filho, "MorphoSys: an integrated reconfigurable system for data-parallel and computation-intensive applications". IEEE Trans. Comput. **49**(5), 465–481 (2000)
11. S.C. Goldstein, H. Schmit, M. Moe, M. Budiu, S. Cadambi, R.R. Taylor, R. Laufer, "PipeRench: A Coprocessor for Streaming Multimedia Acceleration", in *Intl. Symp. on Computer Architecture*, 1999
12. B.V. Essen, R. Panda, A. Wood, C. Ebeling, S. Hauck, "Energy-Efficient Specialization of Functional Units a Coarse-Grained Reconfigurable Array", in *Intl. Symp. on FPGAs*, 2011

Chapter 6
Application of Memory-Based Computing

Abstract Memories are typically associated with data storage in computer system. They either store data temporarily (for example by volatile memories such as SRAM, DRAM etc.) or permanently (for example by non-volatile memories such as Flash, magnetic disks etc.). In this chapter we first draw the outline how the embedded memory may be used for computing as well in a time-multiplexed hardware reconfigurable system. In particular, this chapter makes the following key contributions:

- We propose a hardware architecture to utilize a dense 2-D memory array for the purpose of reconfigurable computing. The main idea is to map multiple multi-input multi-output LUTs to the embedded memory array at each compute block and evaluate them over multiple clock cycles. The proposed framework is therefore a spatio-temporal framework unlike the conventional hardware reconfigurable frameworks which are fully spatial. In the proposed hardware architecture, multiple computing elements communicate with each other over a time-multiplexed programmable interconnect.
- Along with the hardware architecture we outline an effective software framework which maps an input application to the proposed hardware reconfigurable framework. The software flow outlined partitions the input application and maps it to multiple compute blocks, and finally places and routes the mapped design.
- We describe the application domains of the proposed memory based computing model. We identify that the model can be applied to realize a stand-alone reconfigurable framework for mapping random logic as well as it can be used for hardware acceleration of algorithmic tasks.

We believe that the proposed MBC model can be used for computing in diverse contexts for a variety of application domains. Following are some of the application areas where the effectiveness of the MBC framework has already been validated. Other areas are topics of future research.

6.1 Motifs

The proposed memory based computing framework may be used for the purpose of: (i) improving the performance; (ii) energy and/or (iii) reliability of the computing system.

- Interconnect performance and I/O bottleneck are the major limiters to performance improvement in modern computer systems [1, 2]. They have poor technology scalability which will only degrade the performance in future computing systems. A memory based computing framework can be used to improve performance by: (a) computing the data in close proximity with the memory array in which it is stored, thereby mitigating the I/O bottleneck; (b) advocating a temporal computing model at each compute node, it can minimize the requirement for programmable interconnect.
- Similar to performance, energy overhead primarily comes from the movement of data from and to the memory. In MBC, computing is performed closest to the location of data storage and is therefore much effective in reducing the energy requirement for the computing system.
- In nanoscale technologies, both logic and memory are prone to manufacturing defects as well as parametric failures. Redundancy based defect tolerance is a commonly approach in handling such failures. Redundancy in logic however incurs significant overhead in terms of area and power, since the entire logic unit needs to be replicated [3]. Failures in memory can however be easily tolerated by incorporating redundant rows and columns in the memory array and replacing the faulty ones with the functional ones. Hence memory based computing can be effective in improving the reliability of a computing system as well.

6.2 Contexts

The memory based computing model can be applied in diverse contexts. Two specific applications we have already investigated are as follows:

- As a stand alone reconfigurable framework used for LUT based mapping of random logic. We have already demonstrated that mapping the input application into multiple multi-input multi-output LUTs, storing the LUTs into a dense 2-D memory array and evaluating them over multiple clock cycles can be extremely effective in reducing the number of LUTs as well as the programmable interconnect requirement for such a framework.
- As a hardware accelerator it can be used to accelerate both compute and data intensive sections of algorithmic tasks from different domains of applications. Such a framework is hereafter referred to as MAHA—an abbreviation for MAlleable Hardware Accelerator. The idea is to instrument an existing memory organization and augment it with additional logic so that it can be transformed

from a simple storage system to a memory based hardware reconfigurable framework on demand. The chief benefit with such an arrangement is that MAHA can potentially mitigate Von-Neumann bottleneck [4] since it can migrate the application to the data rather than the data to the application. In such framework, the additional logic not only includes control logic for memory based computing but also custom logic for mapping datapath operations such as addition, multiplication etc. It should be noted here that the instrumentation must be performed in a manner that incurs minimal overhead to the memory integration density and performance.
- The third context is to apply the proposed MBC model to commercially available general purpose hardware reconfigurable frameworks such as FPGA. The idea is to map the input application in a spatio-temporal manner on the FPGA itself, with each MLB built in close proximity with the memory in which the data is being stored. This idea is being currently evaluated.

6.3 Domains

Following are the domains of applications which benefit from memory based computing:

- Applications which are rich in complex functions such as trigonometric and transcendental functions as well any other compute intensive functions are amenable to memory based computing. These include applications from diverse domains such as scientific computing, cryptography, image processing, informatics [5] and general purpose signal processing [6].
- Since the proposed MBC framework uses LUTs to map the input application it is amenable for mapping functions with random logic. Applications where simple logic functions can be fused to form complex logical operations which may be mapped as LUTs to the memory array are also attractive for memory based computation.
- Similar to FPGAs, MBC is a generic reconfigurable framework which can map any function. Compared to their logic implementations, operations such as integer and floating point addition and multiplication incur considerable performance and energy overhead when mapped as LUTs. Note that however, for many applications such as signal processing application, the operand width is limited to 16 bits or less [7, 8]. In such scenarios, MBC can be used to substitute multiple datapath operations with a single lookup operation, thus improving performance and energy.

References

1. A. Dehon, "DPGA Utilization and Application", in *Intl. Symp. on FPGAs*, 1996
2. E.S. Chung, P.A. Milder, J.C. Hoe, K. Mai, "Single-Chip Heterogeneous Computing: Does the Future Include Custom Logic, FPGAs, and GPGPUs? in *Intl. Symp. on Microarchitecture*, 2010
3. M. Bushnell, V. Agarwal, "Essentials of Electronic Testing for Digital, Memory and Mixed-Signal VLSI Circuits" (Springer, Heidelberg, 2000)
4. J. Backus, "Can programming style be liberated from the von-neumann style? A functional style and its algebra of programs". ACM Comm. **21**(8), 613–641 (1978)
5. S. Che, J. Li, J.W. Sheaffer, K. Skadron, J. Lach, "Accelerating Compute-Intensive Applications with GPUs and FPGAs", in *Symp. on Application Specific Processor*, 2008
6. R. Tessier, W. Burleson, "Reconfigurable computing for digital signal processing: A survey". J. VLSI Signal Process. Syst. **28**(1/2), (2001)
7. S. Paul, S. Bhunia, "Dynamic transfer of computation to processor cache for yield and reliability improvement". IEEE Trans. Very Large Scale Integrat. Syst., 1368–1379 (2011)
8. J. Dido et al., "A Flexible Floating Point Format for Optimizing Data-Paths and Operators FPGA Based DSPs", in *Intl. Symp. on FPGAs*, 2002

Part III
Hardware Framework

Chapter 7
A Memory Based Generic Reconfigurable Framework

Abstract This chapter describes the hardware architecture for MBC based stand alone reconfigurable computing framework. It first lays down the requirement for a generic reconfigurable framework and how a fully-spatial computing frameworks addresses those requirements. Next it describes the spatio-temporal MBC model and explains how it can be used as a generic reconfigurable framework. The μ-architecture for each computing unit in the MBC framework is then described in detail. Next, it illustrates with examples, how multiple computing units of the MBC framework communicate with each other via a time-multiplexed programmable interconnect. Attention has been particularly given to explaining the synchronization among multiple LUT evaluations mapped to the same or different computing units in the MBC framework.

7.1 A Generic Reconfigurable Framework

A generic reconfigurable framework is one which can map logic functions with arbitrary complexity. It is well-known that with no constraint on area, power and performance, LUT based implementation is capable of mapping any logic function $f(x)$. With "n"-inputs $(x_1 \ldots x_n)$ to a logic function, a LUT with 2^n entries can sufficiently express any logic function for the "n"-inputs. if one wishes to express "m"-output logic function for the same set of "n"-inputs, a LUT of size $2^n \times m$ is required. In the hardware implementation of a generic reconfigurable framework, for a single output logic function of "n"-inputs, the LUT is realized with a one-dimensional memory while for multiple outputs of the same "n"-inputs, the LUT is realized using a two-dimensional memory. From the expression of LUT size, it can be easily observed that the LUT size grows exponentially with the input size "n". Hence a memory which stores these $2^n \times m$ entries becomes prohibitively large even for small values of "n" (16 or more). It is therefore a common practice to partition the logic function $f(x)$ into multiple functions $(f_1 \ldots f_k)$, each with a smaller input set as shown in Fig. 7.1a. Output of some of these functions feed into

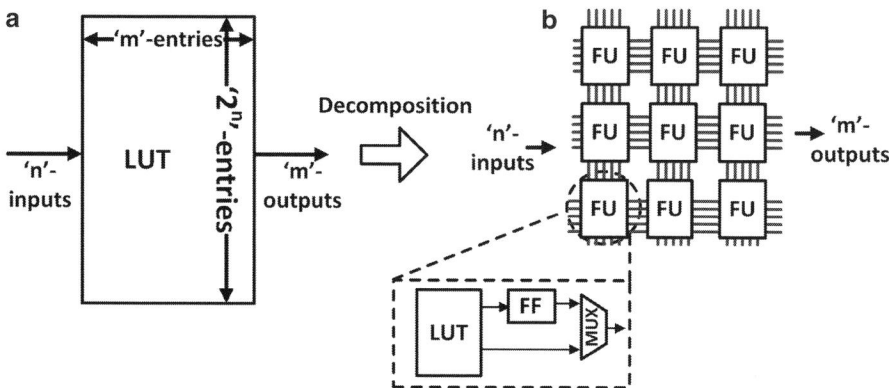

Fig. 7.1 (**a**) A "n"-input "m"-output function can be represented as a LUT; (**b**) A large function can be suitably decomposed into an interconnection of small LUTs and state elements (FF)

other functions, thus building a network of inter-connected functions which when evaluated in a topological fashion produces the same result as the original function $f(x)$. This procedure of splitting a function with a large input set into constituent functions of smaller input set is often referred to as *decomposition*. In a generic reconfigurable framework, each of the smaller functions is mapped to a single LUT and many of these LUTs are inter-connected to each other via a flexible interconnect architecture. Since *decomposition* can produce different interconnection among the LUTs, hypothetically, a generic reconfigurable framework must be capable of supporting any arbitrary interconnection among the LUTs. This is often impractical to implement in the real world. Reconfigurable hardware designs thus often settle for a less flexible solution or pay a large area, delay and energy overhead to facilitate communication between any two LUTs mapped to any two memories in the hardware design. The idea is illustrated in Fig. 7.1b. Since LUTs are expected to behave as "read-only memories", their functionality is pre-defined once a logic function is mapped to this generic reconfigurable framework. Such a framework would work correctly, provided the logic function is stateless. To account for logic functions which have data dependant state transitions, some memory which can be repeatedly updated must be part of this generic reconfigurable framework. From Fig. 7.1b, note that in addition to LUTs, each functional unit (FU) of this generic framework comprises of a sequential element, such as a "flip-flop". Early FPGAs resembled this generic reconfigurable framework [1]. To improve the area, power, performance and resource utilization for such architecture, a number of design innovations have been suggested over years. However, a distributed LUT based architecture still continues to be the underlying generic compute fabric for modern FPGAs.

7.2 A Generic Reconfigurable Framework with MBC

To map any arbitrary function to a MBC framework, the framework also needs to incorporate a sea of LUTs, supported by a flexible interconnection network. However, the critical distinction between MBC and FPGA is that unlike FPGA, where the LUTs are completely distributed in space, in MBC, LUTs are partially localized and partially distributed. The localized LUTs are mapped to a dense local 2-D memory array at each functional unit. In this architecture, local LUTs communicate via a local and flexible interconnect, while LUTs in different FUs communicate via a shared programmable interconnect. In each FU, multiple LUT are executed in a temporal fashion (over time) in a topological fashion and communicates with other FUs (in a spatial fashion) only if the function mapped requires to do so. This spatio-temporal model of MBC allows the user to trade-off locality (resource utilization) against distributed execution (throughput) in an energy-efficient manner. In this section, the generic LUT based architecture for MBC is described in detail. Trade-offs involving incorporating specialized datapath units such as adder or multiple etc. are discussed in following sections.

7.2.1 μ-Architecture for Memory Logic Block

Each compute block of the MBC framework is referred to as the Memory Logic Block or MLB. Each MLB is responsible for topologically executing the LUTs mapped to the dense 2-D memory array local to each MLB. A LUT is ready to be evaluated at a given clock cycle if all the inputs to the LUT are ready. These inputs can be primary inputs or outputs from other LUTs which constitute the function being decomposed and mapped to the MBC framework. For synchronous (with clock) implementation of MBC framework, one or more LUTs are evaluated in a given MLB in a single clock cycle. A MLB therefore finishes the evaluation of all the LUTs mapped to it in one or more clock cycles. Figure 7.2 illustrates the key building blocks of a MLB. It also shows the data and control flow inside a MLB. Before describing each of the MLB components in detail, it is essential to define the term *Granularity*. Let us say, "Z" represents a variable mapped to a MLB (input to or output from a LUT). For "Z", the term *Granularity* refers to the number of bits of "Z" which may be selected separately. For example, if "Z" is a 8-bit variable and *Granularity=1*, then the MBC architecture allows the user and MBC's application mapping tool to individually access $Z[0], Z[1] .. Z[7]$, i.e. individual bits of "Z". However, if *Granularity=4*, then the architecture only allows a group of 4-bits $Z[0], Z[1], Z[2], Z[3]$ or $Z[4], Z[5], Z[6], Z[7]$ to be accessed together. Note that *Granularity* is a design parameter for MBC and all design parameters discussed below are always expressed in multiples of *Granularity*. Each of the MLB components are described below in detail.

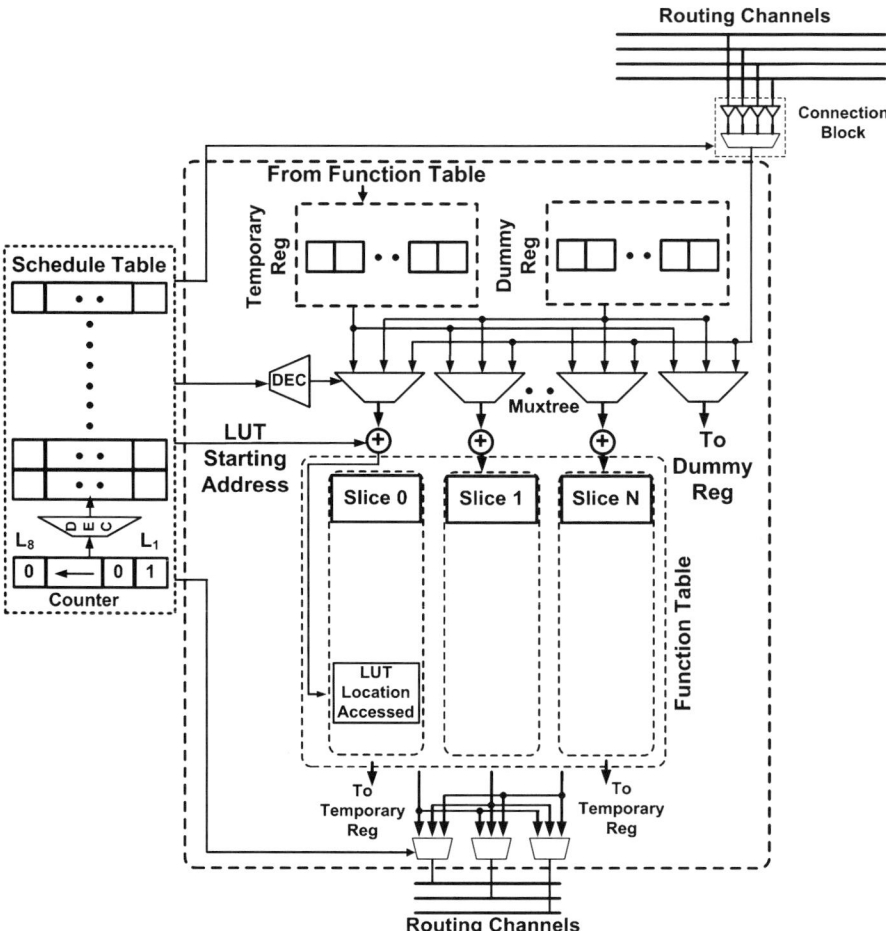

Fig. 7.2 μ-architecture of a single memory logic block (MLB). Figure show the essential hardware components, namely: (i) temporary and dummy registers; (ii) muxtree; (iii) function table and (iv) schedule table

7.2.1.1 Function Table

The principal component of each MLB is the *function table*. It is represented by a dense 2-D memory array to the which multi-input multi-output LUTs are mapped and evaluated over time. The organization of the *function table* is illustrated in Fig. 7.3 and the design parameters which define the *function table* are listed below:

- *Slices:* The implementation of the memory array inside each MLB lends itself to the opportunity of more than one memory access in each clock cycle. This means it is possible to evaluate two or more LUTs in the same clock cycle

7.2 A Generic Reconfigurable Framework with MBC

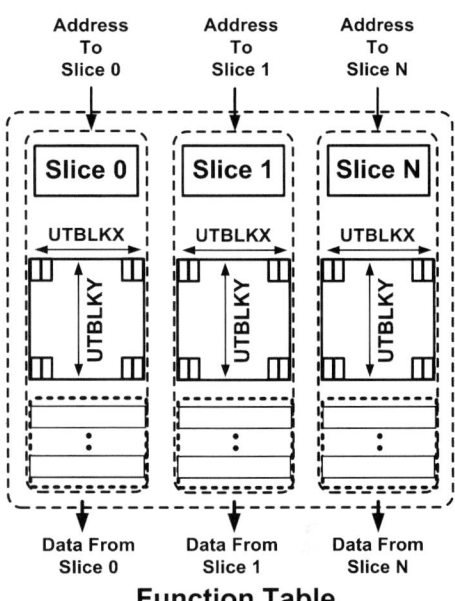

Fig. 7.3 Organization of function table in a MLB

within a given MLB. This is possible if the memory itself is implemented as multiple slices (or arrays), which may be accessed independent to each other. Modern memories provide opportunities for this parallel access either in the form of separate read-port to individual memory cells or by increasing the input bandwidth to the access individual memory arrays in parallel [2]. The MBC model abstracts the underlying memory implementation here and presents a parameterized interface to the mapping tool for application mapping. The maximum number of parallel memory accesses within a given clock cycle inside each MLB is represented by the design parameter *Slices*. Note that however, for # *Slices* parallel memory accesses, the same number of address decoding is required.

- *Rows:* The number of rows in the 2-D memory array used to realize the function table.
- *Cols:* The total number of columns in the 2-D memory array used to realize the *function table*. The total number of memory location in the *function table* is therefore obtained from $Rows \times Cols$. The number of columns for each slice is therefore $Cols/Slices$.

Based on the above design parameters, it is possible to derive the following mapping parameters which facilitate mapping LUTs to the *function table*.

- *Utblky:* If the *function table* supports a total address width of N, then it is possible to access any memory location with an address of bitwidth $M \leq N$. That implies, that *function table* can essentially support any M-feasible LUT where $M \leq N$. In such a scenario, to achieve maximum efficiency in packing the LUTs into the *function table*, the starting address for the LUT accessed needs to be stored beforehand and added with the LUT address to access the actual memory location (refer to Fig. 7.3). To allow the LUTs to start from any location in the *function table* implies the framework needs to store a large starting address for each LUT and add to it the corresponding LUT offset. This degrades both area as well as performance for the MLB. For mapping LUTs, the *function table* has therefore been divided into number of virtual segments both along rows as well as columns. *Utblky* refers to the number of rows inside each segment. A LUT can only start from an *Utblky* boundary. The total number of segments along the rows is therefore *Rows/Utblky*.

- *Utblkx:* Similar to segmentation along the rows, the columns are also divided into virtual segments. *Utblkx* refers to the number of columns in each such segment. It also defines the maximum number of outputs possible from a LUT. *Utblky* and *Utblkx* together define the virtual unit block with which the *function table* can be thought to be composed of.
- *Max_LUT_input:* The number of rows and columns of the *function table* memory array, the number of slices and the *Utblkx* parameter together define the maximum LUT input size possible for a given *function table* design. The formula is given below: $Max\_LUT\_input = log_2(Rows) + log_2(Cols/(Utblkx \times Slices))$.

7.2.1.2 Temporary Register File

At the end of each clock cycle, outputs from the *function table* are latched into the temporary register. At the beginning of the next clock cycle, contents of the register file are read to form the address for LUT access. Design flexibility (local muxtree) allows the contents of any register to form the address for any LUT slice. Two design parameters define the structure of the register file: (i) *Reg_width* defines the width of each register file. The register width should be equal to or greater than the maximum number of output allowed from each LUT slice. (ii) *Reg_count* defines the number of temporary registers inside a MLB. These temporary registers are modeled using the flip-flops since it should allow random bit-level access and the data is expected to be available at the output of the temporary registers as soon as it is written. If a conventional register file implementation is used for the *register file* [3], then care must be taken so as not to overlap read and write operations to the register file.

7.2.1.3 Dummy Registers

In addition to the temporary registers, the MLB architecture currently includes a set of dummy registers to store *feed-forward signals*. By feed-forward signals we mean the signals which are generated in cycle C_i but being used in cycle C_j, where $j > i+1$. Like values stored in temporary registers, these values are also LUT outputs. Since the requirement to read and write into these registers are 1 cycle apart, they may be realized using conventional register file implementation. Their structure is defined by the following parameters: (i) *Dummy_reg_count:* which indicates the number of dummy registers in each MLB and (ii) *Reg_width* which is same as the temporary registers. An alternative to dummy register would be to increase the size of the temporary registers and perform register allocation in such a manner that the feed-forward variables are not overwritten.

7.2.1.4 Schedule Table

The sequence of operations inside each MLB is controlled by the microcode stored inside the *schedule table*. For MBC, the microcode is generated automatically via the application mapping tool. The organization of the *schedule table* is defined by the following parameters: (i) *Maxlvl_per_MLB* which denotes the number of rows in the schedule table and is same as the maximum number of sequential operations that is allowed inside a given MLB. *Maxlvl_per_MLB=K* indicates that operations scheduled inside a MLB will start executing from cycle $C=0$ and continue to clock cycle $C=K-1$, before the operation at clock cycle $C=0$ starts executing again. As shown in Fig. 7.2, a small counter and decoder is present inside each MLB to select the row of the schedule table to read out. The ith row of the schedule table stores the microcode for the ith clock cycle. (ii) *Num_sched_bits* denotes the number of bits in each row of the schedule table. These control bits are responsible for the following inside each MLB: (a) clock gating enable for the temporary registers written in a given clock cycle; (b) clock gating enable for the function table slices which are read in a given clock cycle; (c) starting location (row and column) for the LUTs mapped to a slice of the function table; (d) signals to select the operand source (MLB input, temporary register or the dummy register) for each LUT access to form the LUT address; (e) clock gating enables for the dummy registers to be written in a given clock cycle. Note that the control information can be stored as either as an one-hot encoded information or stored as a compact instruction which can be decoded inside the MLB. The decoded outputs can then be used to control the different operations inside the MLB. Since the schedule table is only written during configuration phase, it can be implemented using any suitable memory element with high read performance/lower read power. A read-optimized register file is a good candidate for the following reasons: (a) the size of the schedule table is much larger compared to the temporary registers. Hence a more compact register file implementation significantly improves the area and energy requirement compared to a flip-flop based implementation of the same. (b) the schedule table is only written during the process of reconfiguration of the MBC platform and is read every clock cycle. Thus there are enough clock cycles between consecutive read and write operations which allows the schedule table to implemented as a conventional SRAM based register file design [3].

7.2.1.5 MLB Input and Output

Design parameters *MLB_DI* and *MLB_DO* denote the number of inputs and outputs to each MLB. Since each MLB operation is scheduled on a clock cycle basis, the MBC application mapping tool exactly knows the inputs and outputs which are available to the MLB in a given clock cycle. Hence the schedule table control bits are programmed in such a manner that in a given clock cycle, the schedule table can select a MLB input to serve as the address for a LUT access. Similarly, if the LUT output in clock cycle $C=i$ from a given MLB becomes the input to another MLB in cycle $C=j | jgeqi$, the schedule table ensures that the LUT output is present at the source MLB output at clock cycle $C=i$. From the source MLB, the LUT value is then routed to the destination MLB via a time-multiplexed programmable interconnect. Note that unlike a FPGA, each input and output of a MLB is time-multiplexed to receive or send different variables over time. In case of FPGA, different values of the same variable is time-multiplexed at each input and output of a compute element.

7.2.1.6 Muxtree

Operations mapped to a single MLB can communicate with each other (outputs of one operation becomes input of another) via a fully-connected local communication network. The communication network is implemented via a tree of multiplexors which can be programmed to route any source signal to any destination. The sources in this case are the temporary registers, the dummy registers and the MLB inputs. Select signals for this local multiplexor tree is derived from the control bits of the schedule table. Design parameters which govern the area, delay and energy for this local programmable interconnect are: (i) *Granularity*; (ii) *MLB_DI*; (iii) *Reg_count*; (iv) *Reg_width*; (v) *Dummy_reg_count*; (vi) *Slices*; (vii) *Max_LUT_input*. It is to be noted that the multiplexor tree serves as a local routing network inside each MLB, and its performance and energy is dominated by the logic delay and energy of the individual multiplexors. This is much unlike the programmable routing resources present in a FPGA, where both performance and energy is dominated by the interconnect parasitics. A transistor level custom implementation (as opposed to synthesis using standard gates) of the multiplexor tree is preferred to its minimize delay, area and performance impact.

7.2.2 Sequence of Operations Inside MLB

Figure 7.4c shows the sequence of events inside a MLB. MLB operations are pipelined with two pipeline stages. The first stage involves fetch of control information from the schedule table and LUT read. The second pipeline stage is operand selection from the MLB input and the temporary/dummy registers.

7.2 A Generic Reconfigurable Framework with MBC

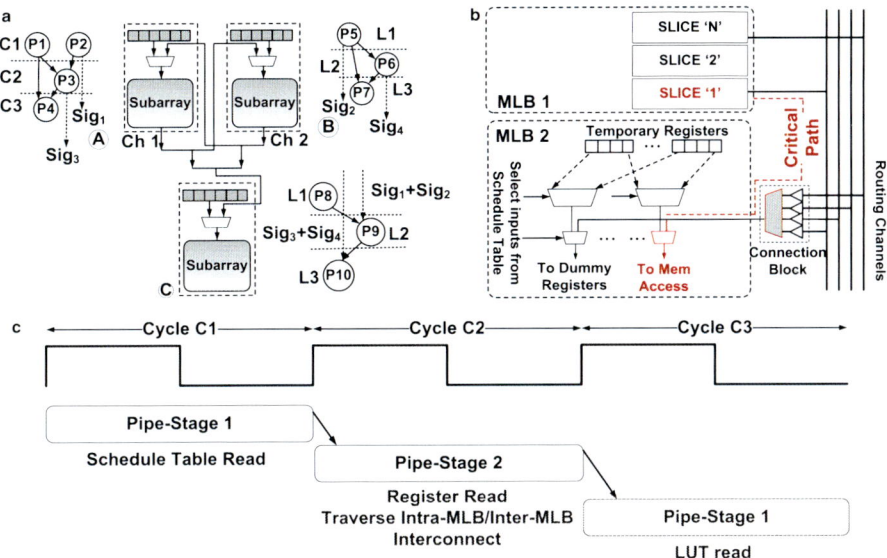

Fig. 7.4 (a) Synchronization of operations across multiple MLBs; (b) Critical path for multi-MLB communication; (c) Sequence of events inside a MLB and across MLBs for execution of applications mapped to multiple MLBs

7.2.2.1 Events in Pipe-Stage 1

For a LUT read which is scheduled in cycle C_i, the schedule table read is performed in cycle C_{i-1} in pipe-stage 1. Control information read from the schedule table serve as control signals for the selection of operands during pipe-stage 2. In pipe-stage 1, the *function table* is accessed in parallel to the *schedule table*. Operands from the register files which form the LUT input are latched using edge triggered flops. For *function table* read, the MBC model does not restrict itself to any specific memory technology. Without loss in generality, it can be assumed that the clock period is long enough to accommodate the 2-D memory array under consideration. Or based on a specific memory read performance, one may choose to insert more pipeline stages in the memory read path to improve throughput. At the end of pipe-stage 1, outputs from the function table are latched into the temporary registers and MLB outputs are selected from the signals produced in the same clock cycle.

7.2.2.2 Events in Pipe-Stage 2

Operands which form a LUT address are selected from MLB inputs, temporary and dummy registers during pipe-stage 2. Thus any LUT operation which is scheduled in cycle C_i will have their operands selected in cycle C_{i-1} in pipe-stage 2. And for that selection, the schedule table is read in cycle C_{i-2} in pipe-stage 1 to obtain the control information. This way, multiple operations are overlapped in time inside a given MLB. The LUT addresses formed from operands serve as offset and is added to the LUT base address stored in the schedule table to form the complete LUT address. Unlike pipe-stage 1, for which the clock period is primarily determined by the memory array access time, the pipe-stage 2 delay is primarily determined primarily by the intra-MLB routing delay through the local multiplexor tree.

7.2.3 Inter-MLB Communication Framework

When multiple MLBs communicate among themselves, the proposed MBC model is similar to other spatial computing frameworks. The major difference is that the same routing channels are used in a time-multiplexed manner, i,e, they carry different signals over multiple cycles. This is in contrast to spatial computing frameworks where interconnects only carry different values of the same signal over time.

Thus, the maximum number of channels that need to be routed to a MLB M_i is $\sum_j (Max(Ch_{i,j,k} \forall k))$, where $Ch_{i,j,k}$ is the number of signals that is transmitted from M_j to M_i in *cycle k*. Figure 7.4a shows the multi-MLB communication framework and synchronization between the operation of multiple MLBs. MLBs *A* and *B* communicate with MLB *C* in *cycle* 2 and 3. The signals that need to be transmitted in *cycle* 2 are *Sig* 1 and *Sig* 3 and those in *cycle* 3 are *Sig* 2 and *Sig* 4, respectively. In spite of the number of signals being 4, only 2 channels (*Ch* 1 and *Ch* 2) are required to connect MLB *A* and *B* with MLB *C*. *Ch* 1 will hold the values of the signals *Sig* 1 and *Sig*2 and *Ch* 2 will hold the values of the signals *Sig* 3 and *Sig*4 in *cycle* 2 and *cycle* 3, respectively. Thus a time-multiplexed operation inside each MLB leads to time sharing of interconnects for inter-MLB communication. Figure 7.4a and c shows the synchronization between the operation of multiple MLBs. All the operations that are scheduled in the same clock cycle in MLB *A*, *B* and *C* are executed in parallel in these MLBs. The inputs to the LUT operations in MLB *C* can come from either internal registers or outputs of MLB *A* or *B* as illustrated in Fig. 7.4a. LUT outputs from MLB *A* and *B* are available before the completion of the clock cycle and routed via the inter-MLB interconnect to MLB *C*. This routing delay is partially overlapped with the pipe-stage 2 operation in MLB *C* which performs register read and operand selection from internal registers in the same duration of time. The worst case delay of the pipe-stage 2 is therefore determined by the maximum of intra-MLB routing delay as well as inter-MLB routing delay. It may happen that an input from cycle C_i is not used in cycle C_{i+1}, instead used in cycle C_k, $k > i+1$. Such a feedforward input would continue to be latched in the dummy registers till cycle C_k. This ensures topological execution of the operations inside each MLB.

7.2.4 Estimation of MLB Cycle Time

The cycle time for an application mapped to a multi-MLB framework is determined by the inter-MLB cycle time ($T_{Inter-MLB}$). Inter-MLB cycle time is different from intra-MLB cycle time $T_{Intra-MLB}$ which is true when MLBs do not need to communicate amongst themselves. While $T_{Intra-MLB}$ purely depends on the delays for the individual pipe stages. The final cycle time after mapping a given function to the MBC framework is however obtained by the maximum of all inter-MLB cycle times. $T_{Inter-MLB}$ is determined by the delay between two successive memory accesses in two communicating MLBs. Since a memory access occurs during pipe-stage 1, the LUT output has one clock-cycle (pipe-stage 2) to propagate from the source MLB LUT output to the destination MLB LUT input. This is dominated by the inter-MLB routing delay ($Delay_{Inter-MLB}$) which stretches the clock period requirement for pipe-stage 2. The expression for clock period during pipe-stage 2 is therefore $Max(Delay_{Inter-MLB}, Delay_{Intra-MLB})$ and the clock period for an application mapped to the MBC framework is evaluated as $T_{MBC} = Max(T_{pipe-stage1}, T_{pipe-satge2})$. The concept is conveyed in Fig. 7.4b.

References

1. J. Rose, R.J. Francis, D. Lewis, P. Chow, "Architecture of field programmable gate arrays: The effect of logic functionality on area efficiency". IEEE J. Solid State Circ. **25**(5), (1990)
2. [Online], "CACTI 5.1". http://www.hpl.hp.com/techreports/2008/HPL-2008-20.html
3. J.H. Tseng, K. Asanovic, "Banked Multiported Register Files for High-Frequency Superscalar Microprocessor", in *Intl. Symp. on Computer Architecture*, 2003

Chapter 8
MAHA Hardware Architecture

Abstract In the previous chapter, the architecture and operation for a MBC based generic reconfigurable framework was discussed in details. In this chapter, the focus is on a malleable hardware accelerator which leverages the memory based computing model for on-demand computation in an existing memory architecture. The accelerator is referred to as *MAHA* which stands for Malleable Hardware Accelerator. The idea is to equip existing memory architectures with additional logic which would allow a storage-only system to be transformed into a hardware reconfigurable platform on demand. This chapter describes the hardware architecture of the MAHA framework and steps for instrumenting an existing memory architecture to realize the same. *Instrumentation* means the design-time modifications which would allow the memory to act as a reconfigurable framework on demand.

8.1 A Typical Memory Organization

CACTI 5.1 documentation [1] presents a detailed description for a typical memory organization. As illustrated in Fig. 8.1, on-chip volatile SRAMs are typically organized in the form of arrays in a hierarchical fashion. The number of levels of in the hierarchy depend on many factors, one of them being the total size of the memory array. Let us consider an organization with 4 levels in the hierarchy. These levels are referred to as *Banks*, *Subbanks*, *Mats* and *Subarrays*. The number of elements in each level of the hierarchy again depend on a number of parameters such as total memory size, width of data input and output bus to the memory array, overhead for the decode logic and also on the constraints (namely area, energy or performance) for which the memory is being designed. As mentioned in [1], the number of elements in each of these levels are design parameters indicated by variables *Banks*, *Ndbl_2*, *Ndwl_2* and *Num_subarrays_per_mat* respectively. *Subarrays* are the elements at the lowest level of the hierarchy and actually represent the 2-D memory array with the associated decoding logic. The bitwidth for the input and output data bus to the entire memory array is denoted by *Buswidth*.

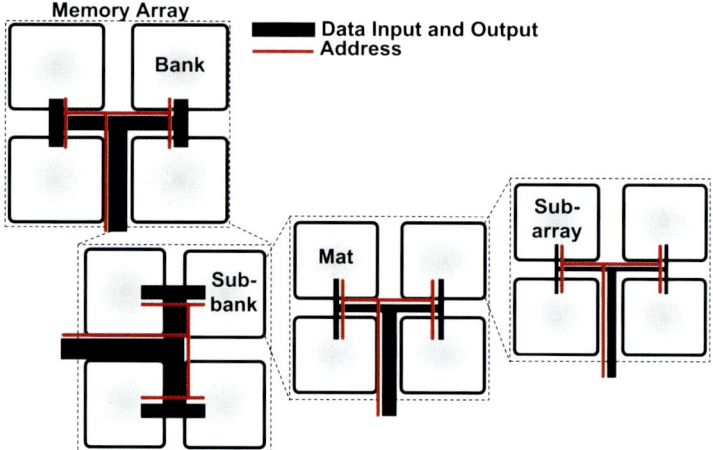

Fig. 8.1 Hierarchical organization of the memory array. The hierarchy consists of *banks*, *subbanks*, *mats* and *subarrays*. The interconnect is also hierarchical where the bandwidth decreases progressively down the hierarchy

Note that the interconnect inside the memory is also distributed in a hierarchical manner with the bandwidth progressively decreasing from the *Banks* to the *Subarrays*. The interconnect follows a hierarchical H-tree pattern [1] where the bandwidth at each level of the hierarchy is given as follows:

- *Num_DI/DO_banks:* Number of data inputs and outputs to each bank. This is same as the *Buswidth* since in memory only one bank can be accessed at one point in time.
- *Num_DI/DO_subbanks:* Data inputs and outputs to a bank gets divided into buses with bitwidth *Num_DI/DO_Subbanks* depending on *Ndbl_2*. The formula is given as *Num_DI/DO_Subbanks = Num_DI/DO_banks/Ndbl_2*.
- *Num_DI/DO_mats:* Data inputs and outputs to a subbank gets divided into buses with bitwidth *Num_DI/DO_mats* depending on *Ndwl_2*. The formula is given as *Num_DI/DO_mats = Num_DI/DO_subbanks/Ndwl_2*.
- *Num_DI/DO_subarrays:* Data inputs and outputs to a mat gets divided into buses with bitwidth *Num_DI/DO_subarrays* depending on *Num_subarrays_per_mat*. The formula is given as *Num_DI/DO_subarrays = Num_DI/DO_mats/Num_subarrays_per_mat*.

The total size of the memory array and the number of elements at level of the hierarchy indicate the number of rows and columns of memory cells present at the lowest level of the hierarchy, i.e. *subarrays*. The design parameters *Row_subarray* and *Col_subarray* define the size for each subarray and the corresponding decode logic.

Apart from the above design parameters, a number of other design parameters define the memory organization. Some of these are: (i) Number of address bits;

(ii) Associativity if the memory array represents the cache data array; (iii) degree of multiplexing at the sense amps etc. During instrumentation, these memory parameters are not modified. As a result, the framework can be seamlessly switched between a reconfigurable computing framework and a normal memory array with minimal impact on normal memory operation.

8.2 Instrumenting Memory for Hardware Acceleration

There are several benefits of instrumenting a memory to function as a reconfigurable framework on demand. It can be particularly useful in scenarios where the data to be processed is already in the memory. In such a case, transforming the memory into a compute framework allows the data to be processed in situ without reading the entire data out from the memory. This offers enormous energy and performance benefit with minimal impact on the normal usage scenario when the memory is being simply used as a storage. There are three key questions to be answered in instrumenting an existing memory organization to perform reconfigurable computing on demand. These are as follows:

- What should be the computing element of such a reconfigurable framework and how should it be realized?
- How can the compute blocks of the reconfigurable framework communicate among themselves?
- How can such framework be seamlessly switched between a normal memory array and a computing framework?

The following text provides answers to the above questions.

8.2.1 MBC for MAHA and Modifications to the MLB μ-Architecture

The major challenge in instrumenting the memory array to perform reconfigurable computing is to add additional logic into the memory array without seriously compromising the integration density of the array. This leaves the designer with two choices:

- Opt for a fully spatial computing architecture similar to the conventional FPGA. This implies the memory arrays should be implemented as small 1-D LUTs, connected to each other through a programmable interconnect framework. In addition to LUTs, datapath elements such as custom adders and multipliers can also be part of this design. It may be noted that a similar organization has been proposed previously in [2]. In [2], the authors propose to model the memory arrays as multi-input multi-output LUTs, connected in a purely spatial manner with multiplexors and dedicated carry chains for implementing up to 2-bit adders and constant coefficient multipliers. The proposed design is used to realize a 2D-DCT transform. However, it can be noted that such a design has several shortcomings when modern memory organization is considered. *Subarray* sizes in modern on-chip memories (such as SRAM array) are typically 8KB or more [1]. For off-chip memories such as DRAM or flash, the minimum array size is much more (256 KB) [3]. It would be extremely inefficient to store the LUTs for 2-bit adders in such large memory arrays and compute them in a spatial manner. Moreover, in a typical memory organization, these arrays do not communicate directly to each other, but only do so with the primary input and output of the memory array through a hierarchical interconnect. Hence it cannot be assumed that the outputs from these subarrays can be cascaded through a programmable interconnect without compromising the memory integration density.

- The other option is to implement a MBC framework on top of the existing memory organization. Although it has several implementation challenges, the reasons in favor of such an option are as follows: (i) MBC is a spatio-temporal framework where each MLB executes multiple operations in a topological manner over multiple clock cycles. These operations are mapped as large multi-input multi-output LUTs to a dense memory array inside each MLB. Hence for the case of MAHA framework, it is possible to construct a MLB with one or more *subarrays* serving as the function table. These subarrays would store the data as well as multi-input multi-output LUTs for the application being mapped to the framework. (ii) Localized temporal computing inside each MLB significantly reduces the demand for PI as required in a fully spatial framework. Hence it can leverage on the hierarchical interconnect already present inside the memory for inter-MLB communication. Clearly the resource utilization is better in case of MAHA.

Figure 8.2 shows the changes made to the existing memory array to realize a multi-MLB framework. As evident from the figure, each of the subarrays have been augmented the additional logic. This additional logic corresponds the control and selection logic for the MLB architecture discussed earlier and includes: (i) *schedule table*, (ii) *temporary and dummy registers*, and (iii) *muxtree*. Note that each subarray acts as the function table inside each MLB. The choice to build a MLB around is each subarray is a design choice based on the maximum area overhead allowed in instrumenting the memory. It is possible to consider a group of N-subarrays as N-slices of the *function table* and build the additional MLB logic around these subarrays. The MAHA framework therefore scales well with the hierarchical organization of current memory arrays. Following modifications are however required at the interface of each subarray to use it for reconfigurable computing:

- In a typical memory organization, the outputs from each subarray in a bank are concatenated to form the bank output and hence the *buswidth*. For typical *buswidth=256* or *512*, and number of subarrays per bank = 8, the bitwidth for each subarray output is 32 or 64. However, the LUTs to which the input

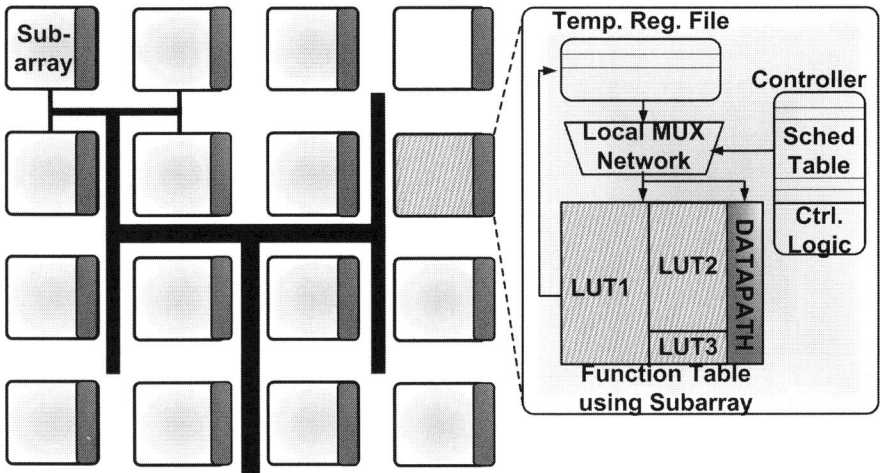

Fig. 8.2 Instrumenting each subarray in the memory organization to construct a MLB. The *subarray* itself acts as the *function table* to which LUTs are mapped

application is mapped typically has 4–8 bit outputs, as defined by the *Utblkx* MLB design parameter. Hence it is necessary to perform an additional level of column multiplexing to select the LUT outputs from the data that is being read out from the subarray.

- During normal memory operation, data is read out from the subarray using the address supplied on the address bits from outside the memory array. However, during operation as a MBC framework, the address to access the LUTs mapped to the memory array are generated from the operands inside the MLB. Hence based on the whether the framework is acting in normal memory mode or as a reconfigurable framework, proper address must be selected to access the subarray.

In contrary to the MBC based generic reconfigurable framework, which maps random logic in terms of LUTs, MAHA is expected to perform as a hardware accelerator engine for accelerating applications from diverse domains. It is well-known that common arithmetic operations such as addition and multiplication and logical operations such as shift/rotate and select operations appear heavily in applications from diverse domains [4]. Many coarse-grained hardware reconfigurable frameworks have therefore been proposed [5–7] where each compute block incorporates the above custom ALUs which perform the above arithmetic and logical operations. The benefit of using custom datapath to perform these arithmetic and logical operations are twofold: (i) performance and (ii) energy. The benefits of using a custom datapath for the above operations has already been validated in the case of FPGA [8]. In order to reap the same advantages, we propose to incorporate

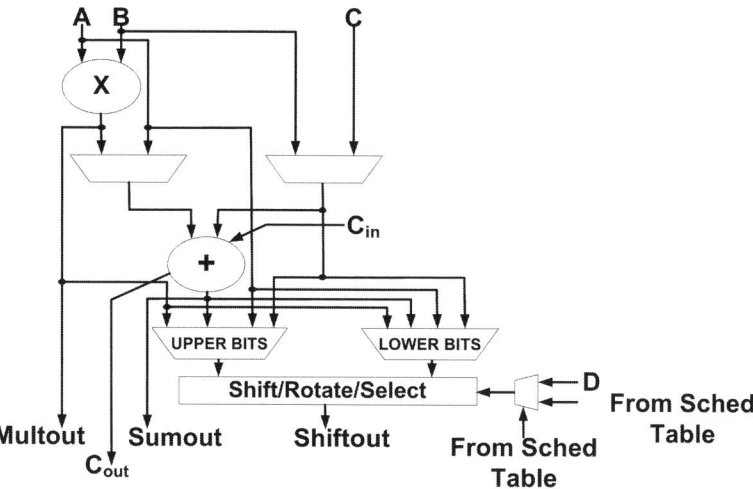

Fig. 8.3 Custom datapath inside each MLB consists of an adder, a multiplier and a shift/select/rotate unit. The size of the multiplier is half the size of the add and shift/select/rotate unit

custom logic supporting (i) addition, (ii) multiplication, (iii) shift, (iv) rotate and (v) select operations inside each MLB. Details on the organization of the custom datapath and its functionality are described below.

Figure 8.3 shows the custom datapath for the MAHA framework. The custom datapath supports a multiplication, an addition and a shift or rotate or select operation in a given clock cycle. Note that to support all the three operations, the inputs must be shared either between the inputs or the outputs of the multiplication or addition operation. For example, the structure can support the following operations concurrently: (i) $A \times B$, (ii) $A + B$ and (iii) $C << / >> shiftamt$ or $A, B << / >> shiftamt$ or $C\ lrot/rrot\ rotamt$ or $A, B\ lrot/rrot\ rotamt$ or $sel?A : B$, where $shiftamt$ and $rotamt$ denotes the amount by which the operands must be shifted or rotated, $lrot$ and $rrot$ denote the left and right rotations and $sel?A : B$ denotes the selection between two-operands based on the select signal sel. Otherwise, the datapath can also support operations such as $((A \times B) + C) << / >> shiftamt$ where the two operands A and B are multiplied and added with another operand C. The result of addition is finally shifted by the $shiftamt$. The same holds true for rotation. The custom datapath as illustrated in Fig. 8.3 is specified by a single design parameter $Addsize$ which denotes the width for the adder. The multiplier size denoted as $Multsize$ is set to half the adder size so that the multiplier output is of size $Addsize$. The bitwidth for the shifter is same as $Addsize$.

Note that the input operand for shift/rotate/select unit is divided into the upper and lower bits to allow the operands A and B to be concatenated before the resultant is shifted or rotated. This allows decomposition of operands of large size using shift/rotate units of smaller size. For multiplication A and B must of size

8.2 Instrumenting Memory for Hardware Acceleration

Addsize/2, while for addition they can be of size *Addsize/2* or *Addsize*. Operand C is always of size *Addsize/2*. Many applications such as Secure hash Algorithm (SHA) require shift/rotate operations of constant value. As shown in Fig. 8.3, this value must be provided from the *schedule table*. The custom datapath operations occur in parallel to *function table* access in the negative half of the clock cycle and each of the multiplier, adder and shift/rotate/select unit output is written into the temporary registers. Select signals for the multiplexors which control the flow of data inside the custom datapath and select the shift/rotate/select signals are read from the schedule table. Each MLB in the MAHA framework with the memory array as the *function table* and the custom datapath thus advocates the same multi-cycle temporal computing model that was proposed for MBC based generic reconfigurable framework.

8.2.2 Multi-MLB Communication in the MAHA Framework

An important challenge in the implementation of the MAHA framework is achieving the multi-MLB communication. The challenge is primarily two-fold: (i) multi-MLB communication must follow the hierarchical interconnect architecture (similar to a H-tree organization) that is already present as part of the memory organization. This is a severe constraint, particularly for applications which are communication intensive; (ii) in case of a stand-alone memory model, the input and outputs for each module in the memory hierarchy comes from the primary input and outputs of the entire memory array. However, in case of MAHA framework, there must be architectural support so that each MLB may receive its inputs from the outputs of other MLBs. In the MAHA architecture the above challenges are addressed in the following manner:

- The first challenge is primarily addressed by the MBC model and the placement and routing algorithm used to place and route the MLBs in the hierarchical memory organization. The temporal computing model inside each MLB significantly reduces the multi-MLB communication requirement as compared to a fully spatial model such as FPGA. Because of the same reason, typical interconnect bandwidth present in the memory array has been found to be sufficient in routing applications as complex as *Motion estimation (ME)* and *Advanced encryption standard (AES)*. The placement and routing procedure followed in our software flow ensures that MLBs which have less communication are spaced furthest apart (in different *banks* or *subbanks*), while MLBs which require more communication are mapped in close proximity (*mats* and *subarrays* within the same *subbank*). With all these techniques, it is possible to satisfy the communication requirement for complex applications using the hierarchical interconnect framework of the baseline memory array.
- The second challenge is addressed by adding programmable switches at the input of each subarray, each mat, each subbank and each bank, such that each

subarray, mat, subbank and subbank can either receive inputs from the outputs of the modules at the same level of hierarchy or from the inputs to that hierarchy level. The idea is illustrated in Fig. 8.4. If there are N-modules (say *subarray*) in a given hierarchy of the memory organization (say *mat*), then the input to *subarray* can come from the outputs of each $N-1$ subarrays and the input to the mat. Now note that the input and output of each *mat* is N times the input and output from each *subarray*. Hence, the switch which selects the input to a *subarray* must be able to select an input from $N-1 + N$ possible sources, or effectively from $2N$ possible sources. The programmable switch used is a simple $2N$to1 multiplexer. The reason for use of multiplexer is to minimize the routing complexity since it offers full connectivity. Typical FPGA frameworks use complex routing switches [9] which allow multi-way connection at the cost of increase in the route complexity. Note that for normal memory read and write operations, the switches at a given level are set in such a manner that it receives the inputs from only the inputs at that particular level.

8.2.3 Modes of Operation

Following modes of operation are identified for a memory which has been instrumented for on-demand reconfigurable computing.

- *Normal memory operation:* Read and write are seamlessly supported to the memory array during this mode of operation. The data written into a particular address must be retrieved during read from the same address.
- *Reconfiguration of the function table:* When the MAHA framework transitions from the normal memory operation to the compute mode, function tables in the MLBs to which the input application is mapped are written first with the LUT contents. This is same as the normal memory write operation with the LUT contents being provided as the *datain* to the memory array. The address for writing the LUTs is also determined during the process of application mapping and provided as address input to the memory array.
- *Reconfiguring the schedule table:* It is essential to reconfigure the schedule table for all MLBs that may be realized on the MAHA framework. For the MLBs which are used to map the input application, the schedule table controls the sequence of operations inside the MLB by storing the appropriate control signals as a microcode. For MLBs which are not utilized, schedule table must store control signals which prevent unwanted datapath and memory operations inside the unused MLB. The data required to reconfigure the schedule table is sent into the memory as part of *datain*. Only in this mode, the *schedule table* is written instead of the *function table*.
- *Reconfiguration of the hierarchical interconnect:* Next the programmable switches inserted in the hierarchical routing network are programmed to allow multi-MLB communication. Values for the configuration bits of these

8.2 Instrumenting Memory for Hardware Acceleration

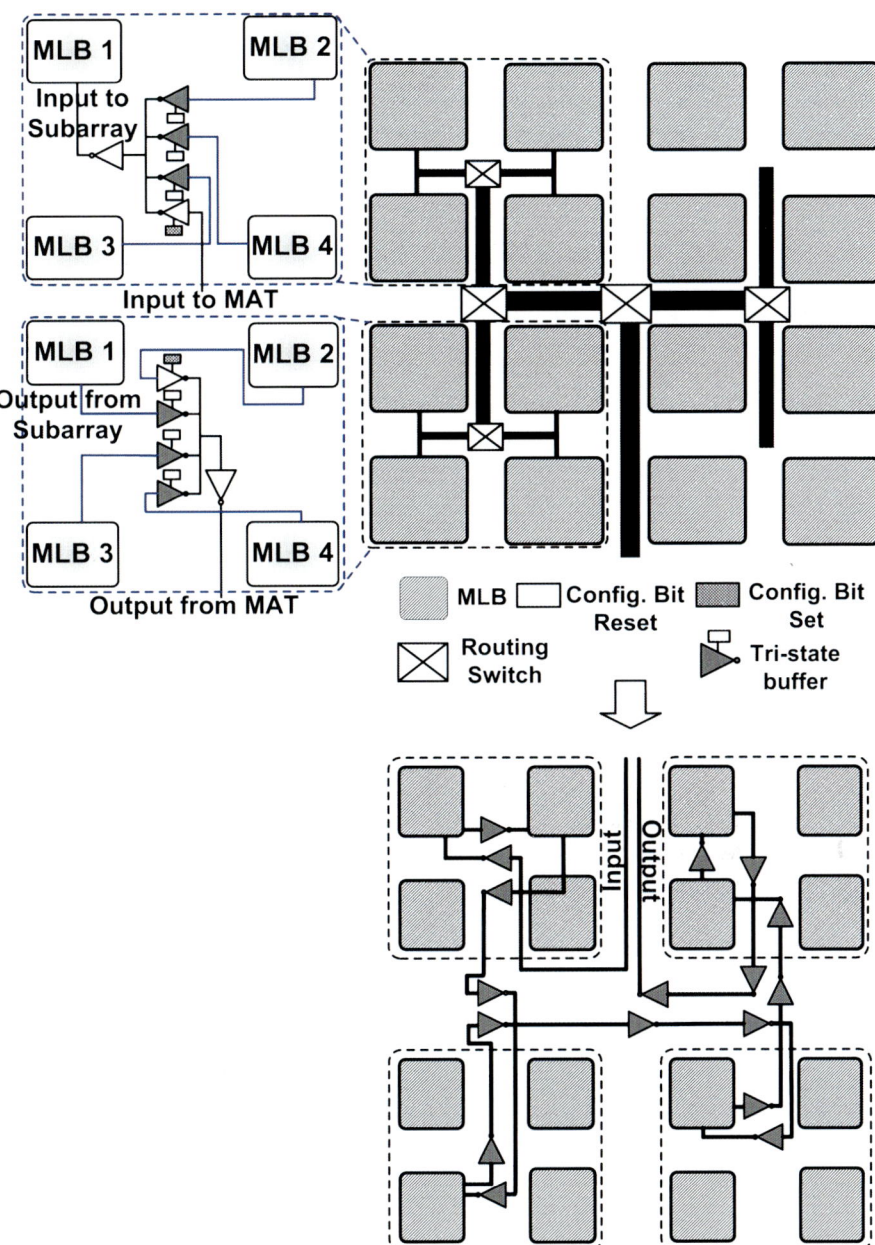

Fig. 8.4 Instrumenting the hierarchical interconnect network in conventional memory organization to support multi-MLB communication

programmable switches are determined once routing is completed. The configuration of the hierarchical interconnect is performed in a hierarchical manner starting with the multiplexors at the input of the banks, going down to subarrays. For programming the routing switches the original memory interface is allowed to have an additional serial programming port. The width of this serial programming interface is defined by the design parameter *Serial_cfg* and is set to 16 bits. Note that this is much smaller than the data input and output buswidth, which is typically 256 bit each. The serial port bitwidth can be reduced at the cost of additional reconfiguration latency. For serial programming, the configuration bits for the switches at each level of the hierarchy are assumed to be are connected in a serial fashion and loaded using the serial data when PI reconfiguration mode is enabled for a particular level in the hierarchy. After the programmable switches are reconfigured, multi-MLB connection is established in the MAHA framework. Following the configuration of the routing switches to provide input to the MLBs, switches which select the individual bank outputs to form *dataout* of the memory array are configured in a similar serial fashion.

- *Execution of the mapped application:* The execution of the mapped execution commences following the configuration of the PI. During this mode, input vectors are provided at the *datain* of the memory array. This input data will be directed towards one or more MLB depending on the mapped application. After execution with the given input vector, primary outputs from one or more MLBs are available at the *dataout* port of the memory. Note that since MBC advocates multi-cycle execution and uses the interconnect in a time-multiplexed manner, for the same input vector set, different input signals are time-multiplexed at the primary inputs for the memory. This is illustrated in Fig. 8.5. Similarly different output signals are present in the primary memory outputs in different clock cycles.
- *Return to memory mode:* Following execution of multiple input vectors on MAHA, the framework returns to normal memory mode by resetting the programmable switches and simply switching the mode control signal.

In order to support the above modes of operation in the MAHA framework, following data and control signals are essential:

1. *Read and write enables:* These enable the read and write operation to the normal memory and also write during *function table* configuration.
2. *cfg_in and cfg_out:* cfg_in denotes the serial port to configure the routing switches at the inputs of each bank, subbank, mat and subarray. *cfg_out* indicates the serial port to configure the switches which select the bank output to form *dataout* of the memory.
3. *mode:* It is the most important control signal which decides one of the 4 modes of operation as described above. The 4 modes are encoded into 2 bits of mode control signal.

8.2 Instrumenting Memory for Hardware Acceleration

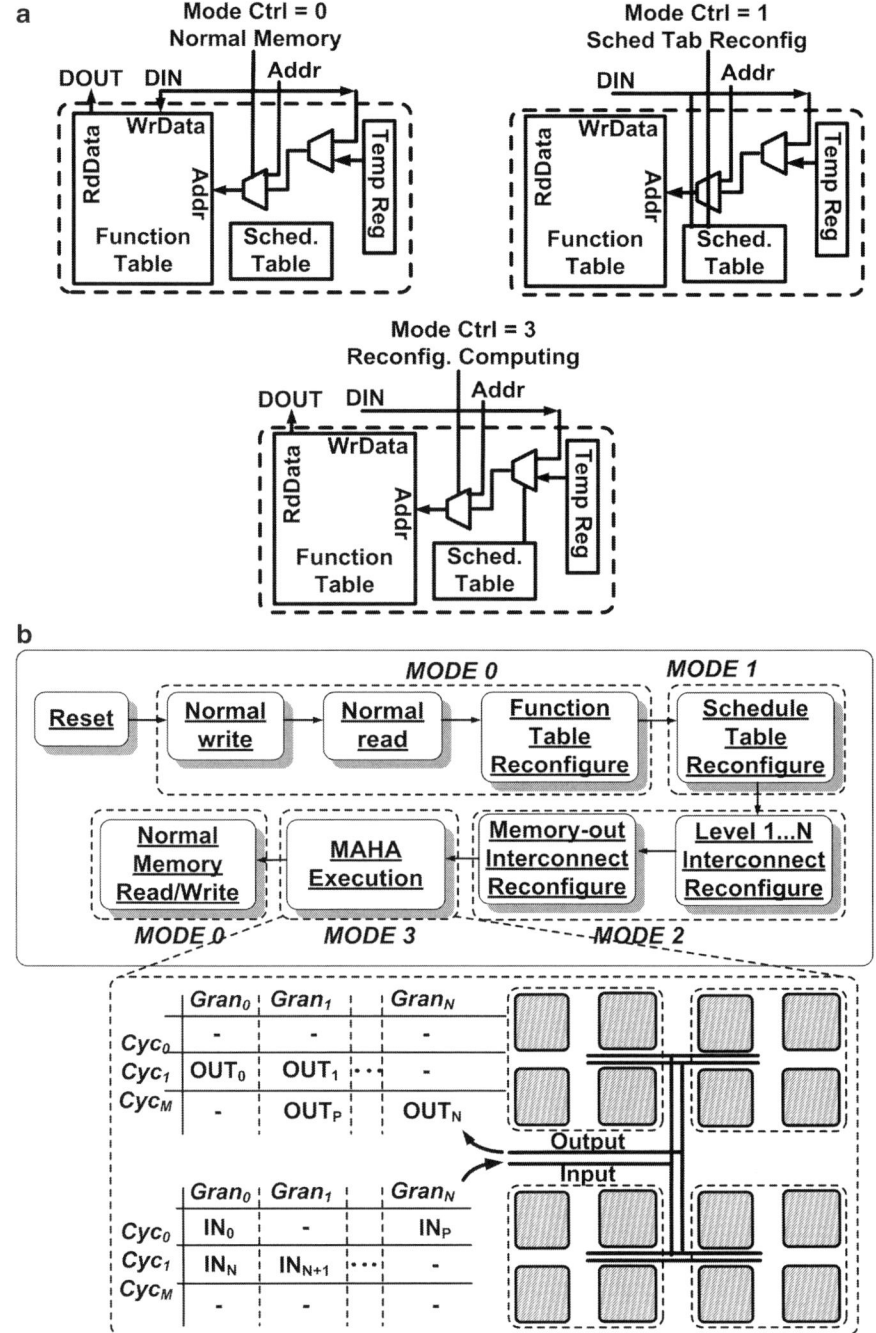

Fig. 8.5 Modes of operation for the MAHA framework. During compute mode inputs must be provided to the framework over the cycles of the cyclic schedule. Outputs are also available from the framework in different cycles of the same schedule

References

1. [Online], "CACTI 5.1". http://www.hpl.hp.com/techreports/2008/HPL-2008-20.html
2. R. Sangireddy, H. Kim, A.K. Somani, "Low-power high-performance reconfigurable computing cache architectures". IEEE Trans. Comput. **53**(10), (2004)
3. K.T. Park et al., "A 45nm 4Gb 3-Dimensional Double-Stacked Multi-Level NAND Flash Memory with Shared Bitline Structure", in *Intl. Solid-State Circuits Conference*, 2008
4. B.V. Essen, R. Panda, A. Wood, C. Ebeling, S. Hauck, "Energy-Efficient Specialization of Functional Units a Coarse-Grained Reconfigurable Array", in *Intl. Symp. on FPGAs*, 2011
5. [Online], "Mosaic Developing power-efficient coarse-grained reconfigurable architectures and tools". http://www.cs.washington.edu/research/projects/lis/www/mosaic/
6. H. Singh, M. Lee, G. Lu, F.J. Kurdahi, N. Bagherzadeh, E.M. Chaves Filho, "MorphoSys: an integrated reconfigurable system for data-parallel and computation-intensive applications". IEEE Trans. Comput. **49**(5), 465–481 (2000)
7. R. Hartenstein, "A Decade of Reconfigurable Computing: A Visionary Retrospective", in *Design, Automation and Test in Europe (DATE)*, 642–649 (2001)
8. I. Kuon, J. Rose, "Measuring the gap between FPGAs and ASICs". IEEE Trans. Comput. Aided Des. Integrat. Circ. Syst. **26**(2), (2007)
9. G. Lemieux, D. Lewis, "Circuit Design of Routing Switches", in *Intl. Symp. on FPGAs*, 2002

Part IV
Software Framework

Chapter 9
Application Analysis

Abstract This chapter describes the frontend of the application mapping flow for the MBC framework. The frontend of the flow accepts an application description in the format of a CDFG. The key step involved in the frontend is partitioning of input application into subtasks (decomposition) to facilitate mapping into the MLB functional units. The other important task is to cluster multiple small tasks into large macro functions which can be suitably mapped to a single functional unit (fusion). While decomposition ensures all input applications can be mapped to the MBC framework, the later step improves performance and energy-efficiency of the framework. Each of these steps involve one or more heuristics which are described in this chapter along with illustrations. The output from the frontend of the tool is a transformed CDFG description of the input application which is then picked up the backend of the software flow for scheduling and resource allocation.

9.1 Application Description Using a CDFG

The input application to be mapped to a MBC framework is specified in a CDFG format, with vertices representing individual operations and interconnections among vertices indicating dataflow among them. For convenience of the user, the application mapping tool has been interfaced with *graphviz* graph visualization software [1] for visually analyzing and debugging the outputs from each intermediate step of the mapping tool. Following are the key steps which together define the application analysis phase. A text file containing the control data flow graph (CDFG) for the input application serves as the input for the application mapping tool. Table 9.1 lists the operation types currently supported by the tool. It is to be noted that while the generic MBC based reconfigurable hardware supports only LUT operations, the *MAHA* architecture supports both LUT as well as various other arithmetic and logical operations as described below. Each instruction/operation is currently represented as a vertex in the CDFG and is specified by the following fields: (i) name, (ii) type, (iii) subtype, (iv) inputs, (v) outputs and (vi) bitwidth. *"name"* field

Table 9.1 Instruction set for MBC framework

Instr Type	Instr Subtype	Inputs	Outputs	Bitwidth	Comment
bitswC	2inadd	a b cin	d e	$x_1\ x_2\ 1$	2-input add bit-sliceable w/ carry
bits	xor	a b	c	$x_1\ x_2$	2-input xor bit-sliceable w/o carry
mult	mult	a b	prod	$x_1\ x_2$	2-input mult, subtype with rand is ignored
delay	rand	a	a_d	x_1	Eqvt to one cycle delay
shift	left/right	a shftamt	a_s	$x_1\ x_2$	Shift operand left/right w/ shftamt
rot	left/right	a rotamt	a_r	x	Rotate operand left/right w/ rotamt
sel	rand	a ·· sel	out	$x_0 \cdots x_n$	Select one from n operands
complex	rand	a ··	lutout	$x_1\ x_2\ x_3$	Arbitrary LUT operation
load	x_n	addr	val	x_1	Load val at addr from memory
store	x_n	addr val		$x_1\ x_2$	Store val at addr into memory

should specify the name for each vertex and should be unique. *"type"* specifies the type of operations, as mentioned in Table 9.1. Each *type* of operation may have several *subtypes*. For example, addition and subtraction both fall under the greater type of operations which can be bit-sliced with a carry function (denoted as *bitswC*). *"rand"* subtypes indicates a particular *type* does not have any *subtype*. For memory operations such as *load* and *store*, the subtype represents the bitwidth of the operand that is read out or stored in the memory. Maximum number of inputs as expected for each type of operation is fixed, except for *select* type which can represent Nto1 selection, where N is variable. The maximum number of outputs for each type of operation is fixed. Bitwidths for each operation indicate the bitwidths of the input operands. Output bitwidth is either implied by the bitwidth of the computing resource or specified in the CDFG when they are inputs to other operations. There is no restriction on the bitwidth, i.e. the size of each input operand. Following are operation types that are currently supported by the software architecture:

1. *bitswC* represents operations which are bit-sliceable with carry can be bit-sliced into sub-operations with smaller bitwidth and these sub-operations can be evaluated in consecutive clock cycles. For example, a 32-bit add operation can be bit-sliced into 4 8-bit add operations, with the carry-input for each 8-bit add coming from the carry-out of the previous 8-bit addition operation. Apart

from addition and subtraction, comparison of two operands is another example operation which can be bit-sliced with a carry output. It has three inputs, two operands and one carry in.
2. *bits* represents operations which are bit-sliceable without carry. Simple logic operations or a combination of them constitute these type of operations. After bit-slicing, the sub-operations can be either scheduled in parallel to each other or in consecutive cycles depending on the resource available.
3. *mult* represents general 2-input multiplication operation. The bitwidth of the product is the sum of the two input operands.
4. *delay* represents one clock cycle delay. It is similar to a delay element in logic design and can be used to create finite state machines. It has a single operand.
5. *shift and rotate* has one operand and one shift/rotate amount. The shift/rotate amount should be atleast $log_2(op_{bitwidth})$, $op_{bitwidth}$ being the operand bitwidth.
6. *sel* represents an N-input to 1-output selection operation. In addition to the N-input operands, it should have one select signal of bitwidth atleast $log_2(N)$.
7. *complex* represents the general class of LUT operations. The definition of the LUT operation is specified by the subtype in terms of the input operands. For the example presented in Table 9.1, the LUT function is specified in terms of the input operands a, b and c. The number of inputs (N) and the bitwidths of each input ($bitwidth_i$) is constrained by the total input size for the LUT.
8. *load, store* are memory read and write operations respectively. The first input represents the address for load/store. For the case of store, the second input represents the operand to be stored in the memory. There is no restriction on the size of the address or the data to be stored. The mapping routine decomposes the data to be load/stored to appropriate number of memory locations determined by the size of the address bits.

With the CDFG specified in terms of the above instruction set, the proposed application mapping methodology as illustrated in Fig. 9.1 maps the input application to the MBC framework. The essential steps of the software flow are described below.

9.2 Decomposition

Decomposition is the process of replacing a vertex with large number of inputs and outputs into multiple vertices with same type and subtype satisfying the input/output constraint for the LUTs as well for the custom datapath. Decomposition routines catering to a wide range of logic and arithmetic functions as well complex transcendental functions [2] is incorporated in the software flow. Details of each decomposition routine are provided below.

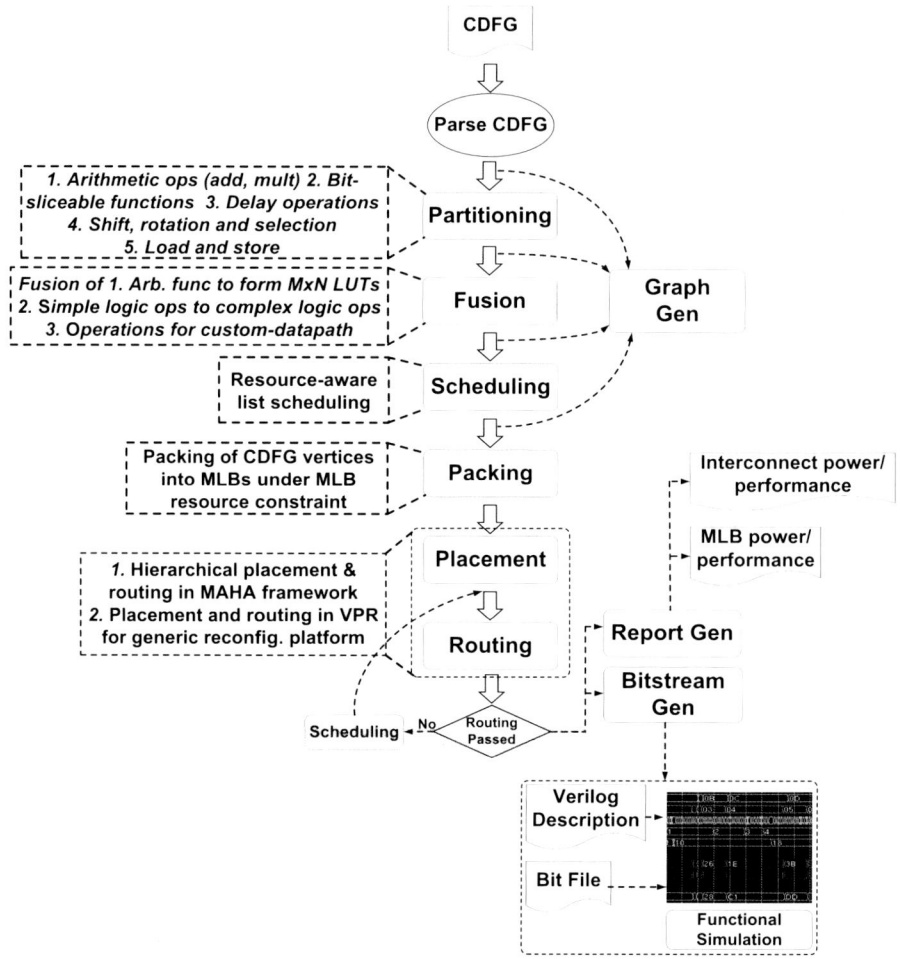

Fig. 9.1 Software flow for the proposed memory based computing model

9.2.1 Decompose_bitswC

For additions and subtractions, this routine decomposes bit-sliceable with carry operations to satisfy the maximum LUT input and output constraint if custom adder is absent. If custom adder is present inside MLB, then decomposition is performed to form sub-operations which fit into the adder size. For comparison operations, the decomposition is always with respect to the maximum LUT input and output values. For example with a 8-b adder in the MLB architecture, a 32-bit addition operation is decomposed into 4 8-b additions. Using *graphviz*, Fig. 9.2 shows the decomposition of a 32-b addition operation.

9.2 Decomposition

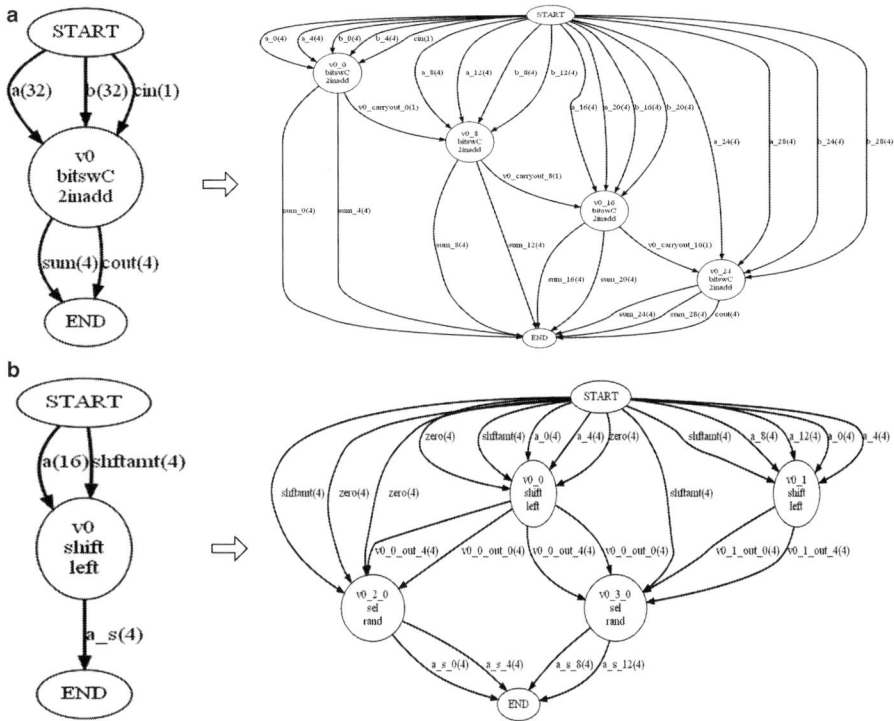

Fig. 9.2 (**a**) A 32-bit addition operation is decomposed into 4 8-bit addition operation for 8-bit custom datapath. (**b**) A 16-b shift operation decomposed to fit 8-fit shifter and (**c**) decomposition of load operation with 16-b address distributed to MLBs with 4KB memory

9.2.2 Decompose_bits

Bit-sliceable operations without carry are decomposed into parallel sub-operations which satisfy the maximum LUT input and output constraint.

9.2.3 Decompose_mult

This decomposes 2-input multiplication into sub-operations which consists of multiplications of smaller bitwidth and partial product addition. The decomposition is performed in keeping with the multiplier size if a custom multiplier is present inside a MLB or according to the LUT input and output bounds.

9.2.4 Decompose_sel

If a custom selection logic is present inside a MLB, a N-input selection operation is decomposed into multiple 2-1 selection. Note the decomposition is performed along the number of operands as well as along the size of the input operands.

9.2.5 Decompose_shift/rotate

Shift/rotate operations are also decomposed into shift and rotate operations of smaller bitwidth according to the size of the custom shift/rotate logic present inside each MLB. The decomposition of the shift and rotate operations are performed following the policy outlined in [3]. The idea is to divide the operand to be shifted or rotated into overlapping segments with each segment equal to the size of the shift/rotate unit. Final shifted/rotated value is selected from the output of these individual shift/rotate units. The idea is conveyed in Fig. 9.2.

9.2.6 Decompose_load/store

As pointed out in [4], the fundamental obstacle to FPGA-based computing today is the FPGA's lack of a common, scalable memory architecture. When developing applications for FPGAs, designers are often directly responsible for crafting the application-specific infrastructure logic that manages and transports data to and from the processing kernels. It is in this respect that an effort was made in [4] to abstract the memory resources present inside the FPGA to mimic a virtual memory system similar to the processor. The *loads* and *stores* operation as defined by our instruction set architecture (ISA) is also meant to serve the same purpose. The proposed decomposition of these operations into individual address spaces of the memories distributed among multiple MLBs is however quite different from the approach proposed in [4]. In [4], the distributed memory inside the FPGA is divided into groups and each group is served by a network-on-chip router which is responsible for matching the address for that group and storing/forwarding data from that group. Abstracting the explicit data distribution and access operations in the above manner is beneficial when the memory under consideration is large. However, the total dataset for many applications (such as discrete cosine transform (DCT), advanced encryption standard (AES) etc.) can be suitably divided into smaller non-overlapping sets which can be operated on in parallel by multiple instantiations of these applications. For these smaller datasets, the proposed decomposition of *load* and *store* would be more beneficial.

The idea is to first allocate memory in one or more MLBs depending on the address size used for load/store and the amount of memory present inside each

MLB. For *load* operation, data from these distributed memories will be read using the lower address bits, while the higher address bits will be used to multiplex the data coming from these distributed memories. For *store*, the address is decoded and only a particular MLB is selected for storing the input data. This decomposition requires only a multiplexer for *load* and a decoder for the *store* operation. The user is still abstracted from explicitly instantiating the memory segments and accessing/storing the data from/to the correct one.

9.3 Fusion

Fusion involves opportunistic reduction in the total number of operations after decomposition (represented as vertices in the CDFG generated after decomposition) by combining multiple operations into a single operation which can be suitably mapped as a LUT or custom datapath operation. It incorporates three routines: (i) fusion of random LUT based operations; (ii) fusion of bit-sliceable operations and (iii) fusion of custom-datapath operations. Details of these individual routines are described below.

9.3.1 Fusion of Random LUT Based Operations

Fusion of random LUT operations is essentially clustering of closely connected vertices which represent LUT operations, such that the input/output constraints for LUTs is not violated. Conventional hypergraph based partitioning techniques [5] however cannot be used to define the partitions, since it can lead to the formation of cycles, which may result in multiple evaluation of the same partition in the MBC framework. In the example shown in Fig. 9.3a, the original netlist is technology mapped to 4-input LUTs. In FPGA, partition P1 can be evaluated twice depending on the relative latency of the inputs $I_7 \cdots I_{12}$. Given enough time a spatial framework such as FPGA will evaluate the output correctly. However, MBC framework imposes the constraint that the original netlist be partitioned in a topological manner and the partitions are evaluated topologically. In light of the above observation, a heuristic for partitioning the target application into multi-input multi-output partitions has been developed. The vertices inside each partition are then fused to form a single vertex to be mapped as a LUT operation.

Pseudocode for the proposed fusion (also referred to as *partitioning* since vertices are first grouped into partitions and thereafter fused) heuristic is presented in Algorithm 1. The input to our partitioning algorithm is a Directed Acyclic Graph (DAG) representation obtained from the decomposition step. Let us denote this graph by $G(V, E)$, where V is the set of combinational nodes and E represents the dependencies between the nodes. The partitioning algorithm begins by topologically sorting the vertices in G, which is accomplished through the following tasks.

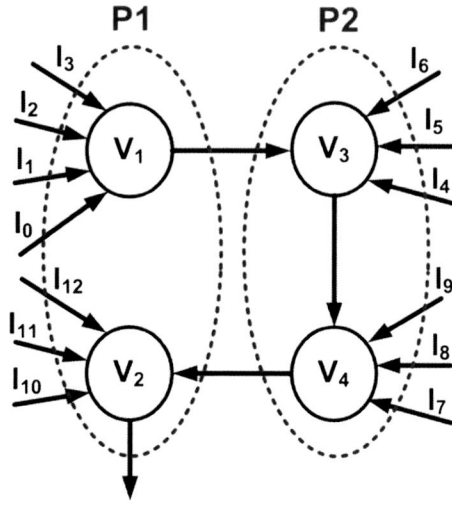

Fig. 9.3 Formation of cycle among partitions in conventional partitioning approaches for FPGA. A similar cycle in the proposed framework would necessitate multiple execution of the same partition

---

Algorithm 1 *Partitioning*

1: **Input:** Circuit as DAG ($G(V,E)$), partition size ($M \times N$)
2: **Output:** Set of partitions (P'')
3: Levelize the vertices $v_i \in V$
4: Sort vertices topologically and store them in V'
5: **for** $i = 0$ to $|V'| - 1$ **do**
6:    **if** $v_i \in V'$ is unmarked **then**
7:       **if** $|P|$ **then**
8:          **for** $j = 0$ to $|P| - 1$ **do**
9:             $temp_j = N_j \bigcup v_i$
10:            **if** $MFFS(temp_j)$ is $\leq$ Max_LUT_input & number of outputs $\leq$ Utblkx & does not form cycle with other partitions **then**
11:                $C = C \bigcup MFFS_j$
12:            **end if**
13:          **end for**
14:          **if** $C \neq \emptyset$ **then**
15:             Find $C_j \ni I(C_j) = min(I(C_i)|C_i \in C)$
16:             For partition P_k corresponding to C_j
17:             Update nodeset of $P_k = C_j$
18:             Mark $v_j | v_j \in C_j$
19:          **else**
20:             Form new partition $p \ni N(p) = MFFC(v_i)$
21:             $P = P \bigcup p$
22:             Mark $v_j | v_j \in MFFC(v_i)$
23:          **end if**
24:       **else**
25:          Form new partition $p \ni N(p) = MFFC(v_i)$
26:          $P = p$
27:          Mark $v_j | v_j \in MFFC(v_i)$
28:       **end if**
29:    **end if**
30: **end for**

- Assign to each vertex $v \in V$ a level $l(v)$ where $l(v) = max(l(v')|v' \in D(v)) + 1$, where $D(v)$ denotes the set of drivers for vertex v. The primary inputs are assigned to level 0.
- Sort the vertices in the descending order of their levels.

Let V' denote the array for the topologically sorted vertices. If there are no pre-existing partitions, a new partition is created with the Maximum Fanout Free Cone (MFFC) [6] with vertex $v \in V'$ serving as the starting node. If there are pre-existing partitions, for each vertex $v \in V'$, the algorithm finds the Maximum Fanout Free Subgraph (MFFS) [6] of the node set $N_i \cup v$, where N_i denotes the set of nodes already present in partition P_i. Let us denote it as $MFFS_i$. P_i is considered as a possible candidate to include vertex v if $MFFS_i$ satisfies the following conditions:

1. M-feasible
2. the number of outputs is less than N
3. $MFFS_i$ does not lead to formation of cycle with other partitions.

$MFFS_i$ is thereafter stored in an array C. Once all the existing partitions are investigated, the most suitable candidate is obtained as P_j where $I(P_j) = min(I(C_i)|C_i \in C)$, $I(C_i)$ being the number of inputs to the MFFS stored as element C_i. This ensures that the vertices with the most overlapping input cone are grouped together.

9.3.2 Fusion of Bit-Sliceable Operations

Bit-sliceable operations that cannot be fused together by the above MFFC/MFFS based fusion routine due to LUT input/output constraint, are fused together and then bit-sliced to satisfy the LUT input/output constraint. Since the minimum size of each operand in the framework is defined by the *granularity*, the number of bit-sliceable operations fused (N) through this routine must satisfy the constraint $N \times granularity \leq LUT_{input}$. The fusion of bit-sliceable operations is currently implemented as a greedy heuristic which first levelizes all vertices in the CDFG. Then in an order of decreasing vertex levels, input cones for vertices with type *bits* are traversed for vertices of similar types which contribute to the immediate fanin of the vertex under consideration. If the number of inputs to these two vertices satisfies the above constraint, they are fused together to form a single vertex. Vertex of type *bits* which form the immediate fanin of this fused vertex are again considered for fusion and the heuristic proceeds in this manner. Unlike MFFC/MFFS based algorithms which are optimal, this greedy heuristic fails to identify the opportunity of fusing vertices when there are reconvergent fanouts. The fused vertices are thereafter bits-sliced in accordance to the maximum LUT output constraint. Examples of the fusion is illustrated in Fig. 9.4.

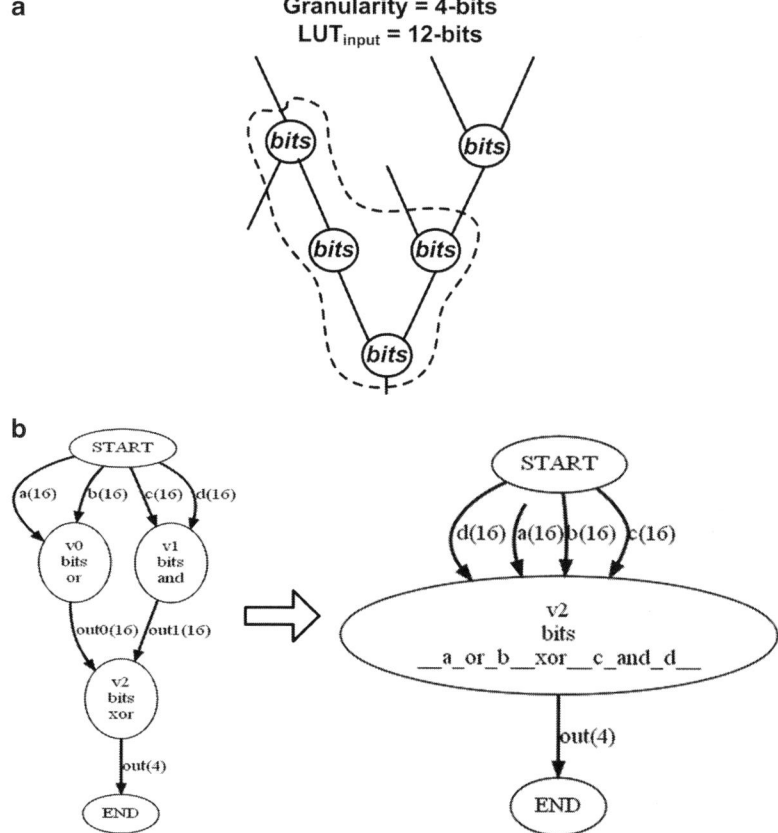

Fig. 9.4 Fusion of multiple bit-sliceable operations into a single bit-sliceable operation

9.3.3 Fusion of Custom Datapath Operations

The custom datapath as described in Fig. 8.3 present inside the MLB lends itself to more opportunities for fusion of individual operations mapped to the custom datapath. For the datapath illustrated in Fig. 8.3, following vertex types can be obtained after fusion:

- *multadds:* stands for the function $(a*b+c) << shiftamt$.
- *multaddsel:* stands for the function $sel?(a*b+c):c$.
- *adds:* stands for the function $(a+b) << shiftamt$.
- *addsel:* stands for the function $sel?(a+b):a$.
- *adds:* stands for the function $(b+c) << shiftamt$.
- *multadd:* stands for the function $a*b+c$.
- *mults:* stands for the function $(a*b) << shiftamt$.
- *multsel:* stands for the function $sel?(a*b):c$.

As evident from the above vertex types, constraint for fusion of the datapath operations is that they must share one or more input operands. Note that operations involving addition are equally applicable for subtraction and shift operation includes both left and right shift.

References

1. [Online], "Graphviz - graph visualization software". www.graphviz.org
2. F.D. Dinechin, A. Tisserand, "Multipartite table methods". IEEE Trans. Comput. **54**(3), 319–330 (2005)
3. "Implementing Barrel Shifters Using Multipliers". http://www.xilinx.com/support/documentation/application_notes/xapp195.p%df
4. E. Chung, J. Hoe, K. Mai, "CoRAM: An In-Fabric Memory Abstraction for FPGA-based Computing", in *Intl. Symposium on Field Programmable Gate Arrays*, 2011
5. G. Karypis, "Multilevel hypergraph partitioning: applications in vlsi domain". IEEE Trans. VLSI **7**(1), 69–79 (1999)
6. J. Cong, Y. Ding, "On Area/Depth Trade-Off in Lut-Based FPGA Technology Mapping", in *Design Automation Conference*, 1993

Chapter 10
Application Mapping to MBC Hardware

Abstract This chapter describes the key steps in mapping the CDFG output after partitioning and fusion to the MBC hardware. A description for the MBC hardware serves as an input to the backend of the software flow. The number of MLB resources and their organization dictate the decisions made by the flow during scheduling, resource allocation, placement and routing. The output from the backend is a bitfile which can be directly loaded into the MBC hardware. The flow is also capable of estimating the power and performance of the input application, provided the power and performance for the individual MLBs and the programable interconnect is provided as an input to the flow.

10.1 Resource-Aware Scheduling

Multiple MLBs in the MBC framework interact in a spatio-temporal manner following a cyclic schedule. For a CDFG representation of the input application, the vertices of the CDFG are executed in a topological manner. It is therefore possible to levelize the vertices in the CDFG, i.e. assign a level to each vertex based on its position in the CDFG. With vertices receiving the primary inputs to the application being in level 0, other vertices can be labeled to be in level i, if it receives the output of any other vertex which is in level $i-1$. Let *Maxlvl_per_MLB* denote the maximum number of sequential operations that can be accommodated in a single MLB. Then following a cyclic scheduling policy, a vertex v in the CDFG with a global vertex level of M maps into the level $M\%Maxlvl\_per\_MLB$ and accommodated inside one of the MLBs. The idea is illustrated in Fig. 10.1 which shows how the vertices are scheduled for the input CDFG shown on the left side. Contrary to purely spatial reconfigurable fabric such as FPGA, the MBC framework allows the application mapping software to easily tradeoff resource usage vs performance by simply specifying the number of custom datapath and memory operations which can be performed in parallel. These are specified by the mapping parameters *Maxcustom* and *MaxLUTaccess* respectively. Since the

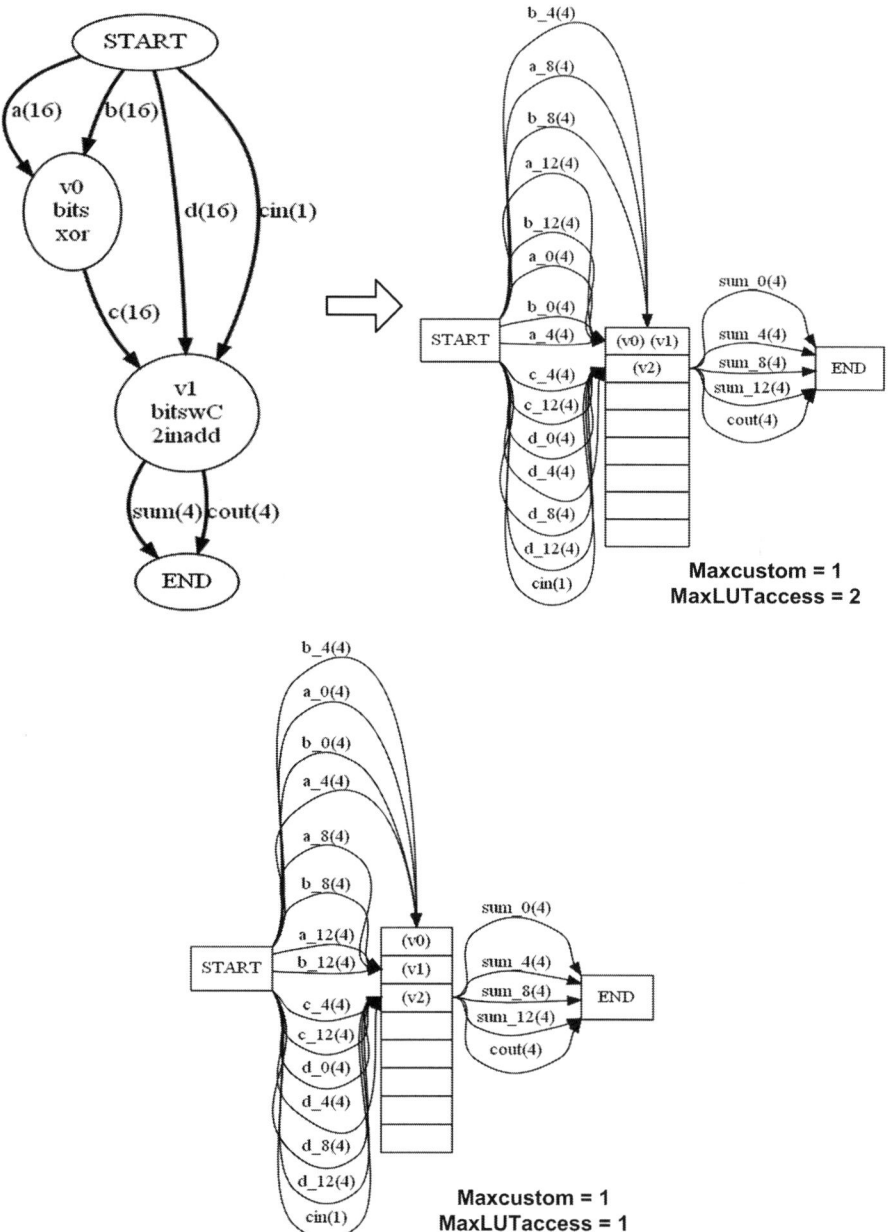

Fig. 10.1 Vertices of an application CDFG after scheduling. For the example maximum number of execution levels inside each MLB is taken to be 8

maximum number of parallel custom datapath and memory access operations are known beforehand, a simple list-scheduling algorithm [1] to schedule the vertices such that maximum resource usage is not violated. From Fig. 10.1, note that for *MaxLUTaccess=2*, it is possible to perform a maximum of two LUT access in parallel in one clock cycle. Thus if the MLB hardware supports two LUT access in parallel, all the *xor* sub-operations can be performed inside a single MLB in a single cycle. The number of cycles however increases by 1, if the parameter *MaxLUTaccess=1*.

10.2 Packing of Partitions to Multi-MLB Framework

Following the resource aware scheduling of the operations denoted by the CDFG vertices, multiple operations are clustered together and assigned to specific MLBs. The heuristics employed in the clustering step try to minimize the inter-MLB communication. This clustering is also referred to as *packing*. However, before the actual *packing* of operations into one or more MLBs, intermediate steps known as *groupLoadStore* is executed. These steps are responsible for grouping decomposed load/store operations into respective MLBs. Following criteria are satisfied while grouping these operations:

- After the load and store operations are decomposed into multiple memory read/write operations, these must be grouped together such that load/store operations pertaining to a range of address must be placed in a single MLB. Moreover, load/store operations for different datasets must be distinguished from each other. A field called *enid* is used for the load/store operations to indicate the dataset to which the load/store operations operate on. After decomposition, if multiple load/store operations are created to manage a single dataset then unique *enids* are assigned to these pairs of smaller load/store operations such that they address the memory within a given MLB. The idea is illustrated in Fig. 10.2.

Following the packing of memory access operations such as loads and stores into one or more MLBs, the packing algorithm ensures that the vertices of other types are packed into the existing or additional MLBs without violating the hardware constraint of individual MLBs. The hardware resource constraints that must be satisfied during the *packing* step are:

- *CheckMem:* Total memory requirement for all the LUT operations and the load/store operations packed inside a MLB must not surpass the total memory available inside each MLB.
- *CheckIO:* The number of signals (considering the bitwidth of the signals) which are inputs and outputs to a MLB in a given clock cycle must be less than the number of the inputs and outputs available to each MLB.

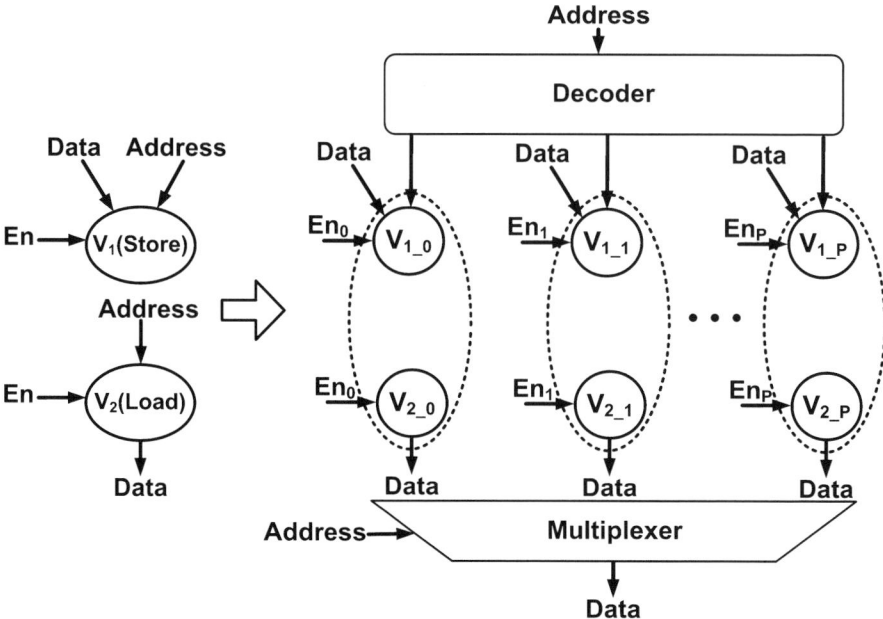

Fig. 10.2 Load and store operations with same enable signals in the CDFG description are decomposed and grouped together in same MLBs

- *CheckBank:* The number of parallel LUT operations inside a MLB in a given clock cycle must not exceed the maximum number of parallel memory accesses allowed in a MLB in a given clock cycle.
- *CheckInterim:* The temporary and the dummy registers present inside a MLB must be sufficient enough to hold all the feedforward signals for the vertices that are packed in a given MLB.

In addition, for MAHA specific software flow, the packing algorithm also ensures:

- *CheckCustom:* The number of custom datapath operations at each level of operation does not exceed the maximum number of custom datapath operations that can be simultaneously executed by the custom datapath. For example, operations $(a+b) << shiftamt$ and $c << shiftamt$ cannot be simultaneously executed within the datapath for a single MLB.

Pseudocode for the proposed packing heuristic is presented in Algorithm 2. A DAG $G'(P,E)$ with partitions (P) obtained from the partitioning step serves as the input to the packing algorithm. PL_i denotes the partition list for a MLB CL_i. The algorithm proceeds by topological sorting of P followed by processing of the partitions in the descending order of their levels. As illustrated in Algorithm 2, the temporary node list $temp_j$ must satisfy the constraints mentioned above before a partition p_i can be included in PL_j. The methodology for calculating the memory requirement

10.2 Packing of Partitions to Multi-MLB Framework

Algorithm 2 *Packing*

1: **Input:** DAG representation $(G'(P,E))$ of the partition set P, Design Parameters (refer to Fig. 11.1)
2: **Output:** Set of MLBs (CL) with the mapped partitions
3: Levelize the partitions $p_i \in P$
4: Sort partitions topologically and store them in P_{sort}
5: **for** $i = 0$ to $|P_{sort}| - 1$ **do**
6:   **if** $p_i \in P_{sort}$ is unmarked **then**
7:     **for** $j = 0$ to $|CL| - 1$ **do**
8:       $temp_j = PL_j \bigcup p_i$
9:       **if** CheckMem($MFFS(temp_j)$)==True **then**
10:         **if** CheckIO($MFFS(temp_j)$)==True **then**
11:           **if** (CheckBank($MFFS(temp_j)$))==True **then**
12:             **if** CheckInterim($MFFS(temp_j)$)==True **then**
13:               $PL_j = PL_j \bigcup p_i$
14:               Mark $p_j | p_j \in temp_j$
15:             **end if**
16:           **end if**
17:         **end if**
18:       **end if**
19:     **end for**
20:     **if** p_i is unmarked **then**
21:       Create a new MLB $m \ni PL(m) = MFFC(p_i)$
22:       $CL = CL \bigcup m$
23:       Mark $p_j | p_j \in MFFC(p_i)$
24:     **end if**
25:   **end if**
26: **end for**

Algorithm 3 *Memory Calculation*

1: **Input:** List of vertices $PL = \bigcup p_i$
2: **Output:** Check for memory violation
3: **for** $i = 0$ to $|PL| - 1$ **do**
4:   $mem_i = 2^{I(p_i)} \times O(p_i)$
5:   $fb = \lceil mem/(Utblky \times Utblkx) \rceil$
6:   $cb = cb + fb$
7: **end for**
8: **if** $cb > Row\_subarray \times Col\_subarray$ **then**
9:   PL violates memory requirement
10: **end if**

---

for $temp_j$ is described in Algorithm 3. Here, $I(p_i)$ and $O(p_i)$ denote the number of inputs and outputs for each partition; fb denotes the number of fine blocks and cb denotes the number of coarse blocks required to map the given partition.

An example of packing operations into multiple MLBs is shown in Fig. 10.3.

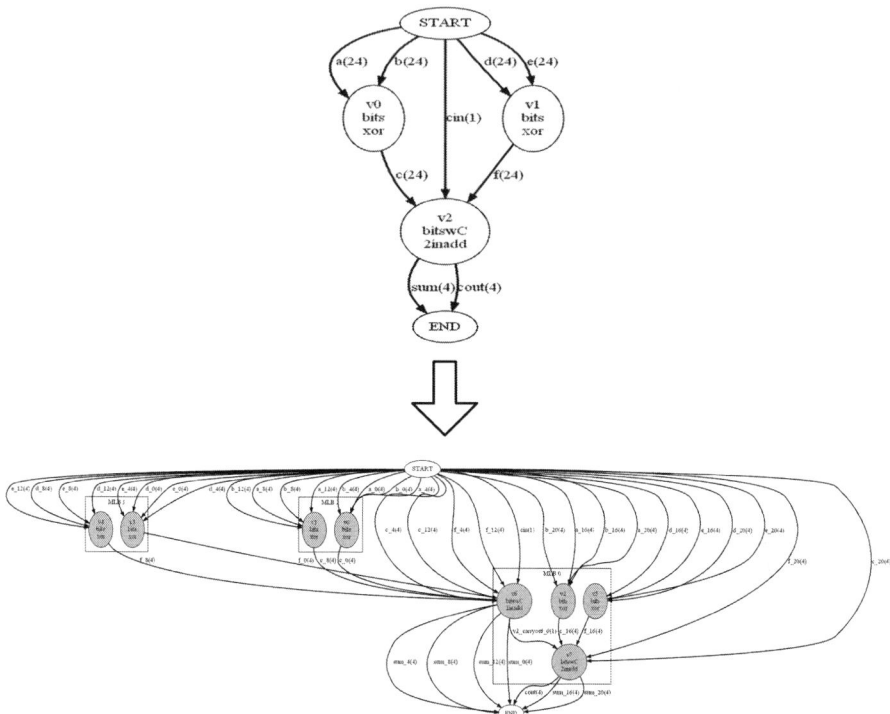

Fig. 10.3 Decomposed and fused vertices of a CDFG are packed into multiple MLBs

10.3 Placement

Following the packing of the operations into multiple MLBs, the MLBs must be placed and routed to ensure correct functionality. Depending on whether MBC is used in a generic reconfigurable hardware or in MAHA for hardware acceleration, the software flow adopts different placement and routing approaches. These are described below.

10.3.1 Placement for Stand-Alone Reconfigurable Computing

A stand-alone reconfigurable computing framework should be able to map applications from diverse domains and therefore comprises of the generic LUTs and fine-grained elaborate programmable interconnect. In modern FPGAs, these LUTs are clustered together in a group of 8–10 [2,3] with both intra-cluster as well as inter-cluster communication. MBC also mimics a clustered FPGA architecture, where the individual LUTs are mapped to the function table. The software flow therefore

10.3 Placement

uses the popular VPR [4] toolset for placing and routing a multi-MLB netlist. The software flow is interfaced with the VPR toolsuite through a *.net* file whose format is compatible with VPR toolset. This file generated from the MBC software flow typically contains the following:

- Description of the inputs and outputs to the LUTs packed into multiple clusters in the FPGA. Clocks and global signals are also distinguished.
- Communication across LUTs inside a cluster.
- Communication across LUTs from one cluster to another.

The *.net* file generated from the MBC flow contains the following information:

- Description of the inputs and outputs of the LUTs packed inside the function table of the different MLBs.
- Intra-MLB communication.
- Inter-MLB communication.

The caveat here is however, VPR does not understand multi-cycle or time-multiplexed operation specific to the MBC framework. To place and route the multi-MLB netlist through VPR, all the operations are therefore unrolled in time and provide an abstraction for time-multiplexed interconnect. This is achieved in the following manner:

- *Modifications to the FPGA Architecture Description File:* Placement and routing of clustered FPGA architecture using the VPR toolsuite requires a FPGA architecture description [5]. This description contains the following important information: (i) position of input and output pins on each cluster; (ii) maximum number of LUTs per cluster; (iii) maximum input size for each LUT; (iv) connection block flexibility (F_c) and switch block flexibility (F_s); (v) resistance and capacitance for the programmable interconnect for inter-cluster communication (vi) sequential and combinational delays inside each cluster and (vii) area of each cluster. In order to adapt the FPGA architecture description to that for the MBC framework following considerations must be made: (i) the position of the input/output pins on the cluster description as well as the values of F_c and F_s are kept the same as in the architecture description file; (ii) since MBC supports variable size LUTs, the maximum number of LUTs that can be mapped to the function table of a MLB is quite large depending on the size of the function table. The maximum number of LUTs inside a MLB is therefore computed based on the minimum LUT size and is provided in the MLB architecture description file; (iii) maximum LUT input size is also modified according to the MLB specifications; (iv) area of cluster is replaced by the area for a a MLB; (v) in comparison to the FPGA architecture, area and delay values for the programmable interconnect are re-calculated to account for the increased area of each MLB tile and hence the increase route length.
- *Modeling the time-multiplexed programmable interconnect:* Let $Input_{i,k}$ and $Ch_{i,j,k}$ denote the number of inputs to MLB_i and the outputs from MLB_i to MLB_j in clock cycle k. In order to model a time-multiplexed programmable

interconnect, the maximum values of $Input_{i,k}$ and $Ch_{i,j,k}$, $\forall k$ is computed and assigned to the respective MLBs in *output.net*. This ensures that only $Max\{Ch_{i,j,k}|\forall k\}$ routing channels are used to route the signals from MLB_i to MLB_j. Similarly the inputs for all the LUTs mapped to a MLB must come from the output of other LUTs mapped to the same MLB or $Max\{Input_{i,k}|\forall k\}$ MLB inputs. Note that through packing it is ensured $MLB\_Input \geq Max\{Input_{i,k}|\forall k\} \geq Max\{Ch_{j,i,k}|\forall k\}$.

The MLB architecture description file and the multi-MLB netlist serve as inputs to the VPR routing tool. Placement and routing using VPR is then performed with no constraint on the maximum number of routing channels.

10.3.2 Placement for MAHA Framework Using a Hierarchical Interconnect

To realize a MAHA framework, on-chip/off-chip memories can be instrumented to create a reconfigurable framework on demand. In the MAHA framework, subarrays are instrumented to create MLBs and the large bandwidth that already exists inside the memory is used for inter-MLB communication. The interconnection network that exists inside the memory is however hierarchical, resembling a H-tree [6]. A hierarchical placement and routing algorithm is therefore incorporated into the software flow to support the MAHA framework.

Given the number of modules in each level of the memory hierarchy and their I/O bandwidth, the flow places the MLBs in a hierarchical fashion such that the number of inputs and outputs crossing each module is minimized. To realize the same, a bi-partitioning approach has been implemented based on Fiduccia–Matheyses algorithm. In this partitioning approach, MLBs are first partitioned among the first level modules. This is followed by distribution among the second level modules and is continued till each MLB has been mapped to the lowermost memory module (a block). The cost function for the bi-partitioning approach is taken to be $\sum_{i=1}^{N} \sum_{j=1}^{N} (-1)^k \frac{b_{i,j}}{l_{i,j}}$, where N denotes the total number of MLBs, $b_{i,j}$ denotes the number of edges between MLB_i and MLB_j and $l_{i,j}$ denote the difference in the level between the source and sink operations inside the two MLBs. $k = 1$ when both MLB_i and MLB_j are inside the same partition of a bi-partitioning step and $k = 0$ when they are in two different partitions. Effectively, two MLBs which communicate infrequently or if the communication is between two operations scheduled far apart in time are placed into different modules.

Routing of primary inputs and outputs to the DFG and the inter-MLB signals is performed in the following hierarchical order:

- Routing of primary inputs to each MLB for all levels of the cyclic schedule.
- Routing of signals which cross the first level of memory hierarchy for all levels of the cyclic schedule.

10.4 Report Generation

- Routing of signals which cross only the second level of memory hierarchy for all levels of the cyclic schedule. This procedure is repeated for each level of the memory hierarchy.
- Routing of primary outputs from each MLB for all levels of the cyclic schedule.

The order simplifies the static switch-box allocation across multiple clock cycles. A signal is routed from the primary input to a memory block on $Channel_i$, only if corresponding channels at upper levels of memory hierarchy is free as well. A channel is considered to be available for routing only if: (i) no other signal is scheduled on the channel in the same clock cycle. (ii) if the switch box which drives the channel is set in a manner such that the source and destination modules are same for the given signal. An already set switchbox indicates it is used to route a signal with the same source and destination for another cycle of the cyclic schedule.

Figure 10.4a summarizes the essential routines of the placement and routing step. In order to address the scenario where the placement or routing fails due to insufficient routing resources, a simultaneous placement-routing-scheduling routine has been incorporated as part of the software flow which re-schedules vertices so as to minimize the routing congestion for a given execution level. During the process of re-scheduling, the execution level with maximum congestion is first identified. Then vertices with most inputs outside the MLB are iteratively moved to the next execution level. The dependant vertices are also shifted from their respective levels. This is continued until the routing succeeds or the vertex returns to the original level in the cyclic schedule. Figure 10.4b shows the placement of the MLBs shown in Fig. 10.3 for a representative hierarchy of banks, subbanks, mats and subarrays. "NULL" denotes empty memory blocks.

10.4 Report Generation

Following the successful placement and routing (P&R) step, reports are generated for estimating the area, power and performance of the mapped design. Similar to P&R, the report generation is also distinct for a MBC based generic reconfigurable framework and a MAHA architecture.

10.4.1 Estimation of Energy and Performance for MBC Based Stand-Alone Reconfigurable Computing

Following are the important steps to the report generation for MBC based stand-alone reconfigurable framework:

- *Area Estimation:* The area of the MLB can be approximated as the sum of the areas of the individual MLB components. The formula is given as:

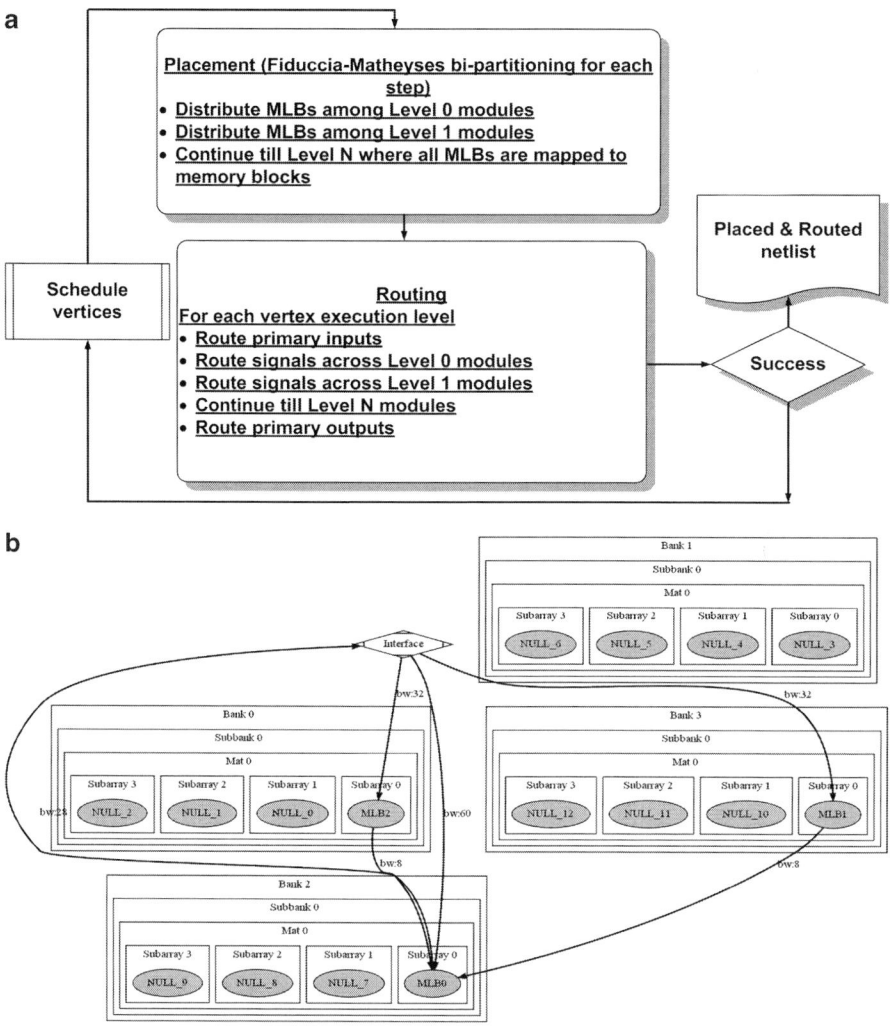

Fig. 10.4 (a) Important steps of the simultaneous scheduling-placement-routing heuristic; (b) MLBs after placement. For the example, the number of modules at each level are 4(Banks), 1(Subbanks), 1(Mats) and 4(Subarrays) respectively

$Area_{MLB} = Area_{controller} + Area_{SchedTable} + Area_{Regfile} + Area_{Dummyreg} + Area_{Muxtree} + Area_{FunctionTable}$.

- *Delay Estimation:* As mentioned earlier, the delay for an application mapped to the MBC framework is a function of both intra-MLB and inter-MLB delays and is calculated as: $Delay_{total} = Max(Delay_{inter-MLB}, Delay_{pipe1}) + Delay_{pipe2}$, where $Delay_{inter-MLB}$ represents the worst case inter-MLB delay after the application is mapped to the MBC framework. $Delay_{pipe1}$ and $Delay_{pipe2}$ represent the critical delays for pipeline stage 1 and 2. $Delay_{pipe2}$ consists of temporary register and dummy register read time and formation of effective addresses to access the function table. This delay is accurately calculated from SPICE simulation. The pipe-stage 1 delay is dominated by the function table access time. Since the function table is modeled as a standard 2-D dense memory array, we have used CACTI [6] in estimating the area, energy and power for different sizes of function table. The worst case inter-MLB delay is obtained from the critical path reported by VPR after placing and routing the multi-MLB netlist. A script was written to parse the MLB to MLB delay in the critical path report and flag the worst case inter-MLB delay. With these three parameters, the total time period for the application mapped is ascertained.

- *Energy Estimation:* For MBC based stand alone reconfigurable framework, currently we only report the MLB energy (dynamic + leakage), i.e. the total energy expended in the MLBs per input vector and do not report the routing energy. The total energy per vector is calculated as: $Energy_{total} = \sum_{i=1}^{N} \sum_{j=1}^{M} (Energy_{controller} + Energy_{Schedtab} + Energy_{Regfile} + Energy_{Dummyreg} + Energy_{Muxtree}) + \sum_{i=1}^{K} Energy_{FunctionTable}$. As evident from the above expression, the total energy per input vector is a function of the total number of MLBs (M) as well as the total number of cycles (N) required to execute a single vector of the mapped application. Energy is expended in the controller, schedule table, register file, dummy register and muxtree in all the MLBs in all clock cycles N. However, the energy expended in the function table is a function of the total number of LUT operations (K) actually required by the mapped application to evaluate a single input vector. In the software flow, the logic energy is calculated through SPICE simulations while the memory energy is obtained from CACTI. The software flow estimates the routing energy by propagating the logic values through the LUTs and across the MLBs via the routing channels.

10.4.2 Estimation of Design Overhead for the MAHA Framework

The area, performance and energy estimates for the MAHA framework are arrived at in a manner similar to the stand-alone reconfigurable framework. Considering the custom datapath present inside each MLB, the area, performance and MLB energy for the mapped application is modified in the following manner.

- The area for the custom datapath is now added to the total MLB area. The schedule table area is also increased in order to store the microcode relevant to custom datapath operation. The number of temporary registers is also increased to store the outputs from the custom datapath. Furthermore, the muxtree also increases in size to select the operands for the custom datapath.

- Since the custom datapath operations occur in pipe-stage 1 in parallel to the function table access, the delay for the custom datapath operations is unlikely to affect the performance in case of MAHA framework.
- The total energy is now augmented by the energy of the operations which are mapped to the custom datapath. The schedule table access energy and the muxtree energy also increases with the size of the custom datapath width.

In addition to single MLB area, multi-MLB performance and total MLB energy, our software flow also reports the following information for the application mapped:

1. Number of vertices in the original CDFG, vertices after partitioning and fusion.
2. Break down of the vertex types.
3. Size of reconfiguration data that is required to on-demand configure the MAHA framework from a simple storage system to the computing platform. This is calculated as $\sum_{i=1}^{M}(Reconfig_{FunctionTable_i}, Reconfig_{SchedTab}) + Reconfig_{PI}$, where $Reconfig_{FunctionTable_i}$ denotes the data required to configure the function table for MLB_i. $Reconfig_{SchedTab}$ denotes the size of the schedule table reconfiguration data and depends only on the schedule table size, irrespective of the MLB number. $Reconfig_{PI}$ denotes the reconfiguration data for the programmable interconnect in MAHA framework.
4. The number of routing switches used at each level of hierarchy in the hierarchical PI framework is also reported in case of MAHA.
5. It reports the total interconnect area on the basis of the number of the total number of elements at each level of hierarchy (banks, subbanks, mats or subarrays) for a hierarchical interconnect.
6. Based on the number of routing switches used at each level of hierarchy, it estimates the total routing energy for each input vector as $\sum_{i=1}^{L}(SwitchCount_i \times Switchenergy_i)$, where L denotes the total number of levels in the hierarchical interconnect, $Switchcount_i$ denotes the number of switch used in level i of the hierarchy and $Switchenergy_i$ denotes the energy per switch for switch at level i.

The software flow summarizes the power and performance for the application mapped to the MAHA architecture in an output report file.

10.5 Bitfile Generation and Functional Validation

The bitfile generation routine is specific to the MAHA software flow. It accepts the placed and routed netlist and performs the following important functions:

- It generates the control or select bits for the programmable switches used to route the signals at different levels of the hierarchical interconnect.
- Since the connection-box both at input and output of a MLB is dynamically scheduled, it generates the schedule table entries which select appropriate inputs from input channels in a given clock cycle. Similarly it generates the microcode to steer the outputs from the function table and the custom datapath to proper output channels.

- In addition to input and output control signals, it also generates schedule table entries which correspond to: (i) enables for the custom datapath and memory operations; (ii) register enables for writing into the register file and (iii) starting location of LUTs being accessed in each clock cycle.
- In addition, it performs the important task of liveliness analysis for intermediate variables and register allocation. If a signal is *live* over multiple clock cycles, i.e. generated in a particular clock cycle but consumed after or over several clock cycles, it moves the signal from the temporary register to an appropriate location in the dummy register so that its value is not overwritten in the following clock cycle.

The bitfile generated by the tool for a given application can be directly loaded into the MAHA verilog model. Functionality in the MAHA mode is then validated by specifying the input vectors for the mapped application.

10.6 Design Space Exploration

The tool allows the user to perform design space exploration of the MBC framework. The user may vary a number of MBC architectural parameters and note their impact on the area, performance and energy for both MAHA as well as the stand-alone reconfigurable framework. These parameters are:

1. Number of elements as well as the interconnect bandwidth at different levels of the hierarchy for the hierarchical interconnect of the MAHA framework.
2. Width of the custom datapath inside each MLB.
3. Number of parallel LUT access inside each MLB.
4. Maximum number of consecutive levels of execution inside a single MLB.
5. Maximum number of parallel custom datapath *(Maxcustom)* and LUT operations *(MaxLUaccess)* allowed for the mapped application.

In order to perform design space exploration, the user is required to:

- Provide an explore file which contains the area, performance and energy values for different MBC components for each design space configuration. For example, if the maximum number of parallel LUT accesses inside a MLB varies as 1, 2 and 4 and the maximum number of consecutive levels of execution varies as 2, 4 and 8, then the explore file must contain the area, performance and energy values for the MLB components corresponding to the configurations 1×2, 1×4, 1×8, 2×2, 2×4, 2×8 and 4×2, 4×4 and 4×8 respectively.

At the end of design space exploration, area, performance and energy for the mapped application corresponding to different design points can be found in a explore report file. The user may then select the best MBC hardware configuration depending on the design constraint.

The software framework for MBC serves several important purposes. These are summarized below:

1. Given a MLB architecture specification, it maps the CDFG for the input application to a multi-MLB netlist.
2. It places and routes the multi-MLB netlist.
3. Given the area, performance and energy estimates for the MLB and the PI, it calculates the total area, energy and performance for the mapped application.
4. Given a set of input applications, it allows the user to vary the MLB architectural parameters and arrive at an optimal MLB configuration for those applications depending on whether the design constraint is area, performance or energy.

References

1. K.K. Parhi, "VLSI Digital Signal Processing Systems: Design and Implementation" John Wiley & Sons (1999)
2. E. Ahmed, J. Rose, "The effect of LUT and cluster size on deep submicron FPGA performance and density". IEEE Trans. Very Large Scale Integrat. Syst., **12**(3), 288–298 (2004)
3. [Online], "Improving FPGA Performance and Area Using an Adaptive Logic Module". www.altera.com/literature/cp/cp-01004.pdf
4. [Online], "VPR and T-VPack 5.0.2 Full CAD Flow for Heterogeneous FPGAs". http://www.eecg.utoronto.ca/vpr/
5. [Online], "iFAR – intelligent FPGA Architecture Repository". http://www.eecg.utoronto.ca/vpr/architectures/
6. [Online], "CACTI 5.1". http://www.hpl.hp.com/techreports/2008/HPL-2008-20.html

Part V
MBC Design Space Exploration

Chapter 11
Design Space Exploration for MBC Based Generic Reconfigurable Computing Framework

Abstract This chapter describes the design space exploration for a MBC based generic reconfigurable framework. Design space exploration identifies the appropriate values for several design parameters of the framework based upon the technology node, the applications being mapped and the power and performance constraints provided to the MBC software flow. To be specific, the exploration phase determines the configuration for both compute blocks (MLBs) and the programmable interconnect. For this exploration, information regarding the variation in design overhead with varying specification for individual hardware components of the MBC framework must be provided to the software flow. The MBC software flow by itself generates an exhaustive combination of all design points ensuring that at each design point a given input application can be mapped to the MBC framework. A mapping result (power and performance) is generates for each of these design points. The user may then select one of the design points based on whether power and performance is more critical to mapping the input application.

11.1 Benchmarks and Experimental Setup

Choice of benchmarks used for arriving at an optimal MBC hardware framework is critical to the final hardware configuration that is being selected. The approach followed here is to first select a set of commonly occurring representative applications (benchmarks) and map them to the MBC based generic reconfigurable framework for a given choice of design parameters. The effect on the geometric mean of performance and energy across all the benchmarks is then noted. One or more design parameters are then modified and the effect is noted again. This procedure is continued till the performance or energy estimates become unacceptable. MCNC and ISCAS'85 and ISCAS'89 benchmarks [1] were selected for validating the effectiveness of the MBC based generic reconfigurable computing framework.

Fig. 11.1 MLB design parameters for MBC based generic reconfigurable computing and additional design parameters for MAHA. Scheduling parameters allows the designer to trade-off between performance and resource usage

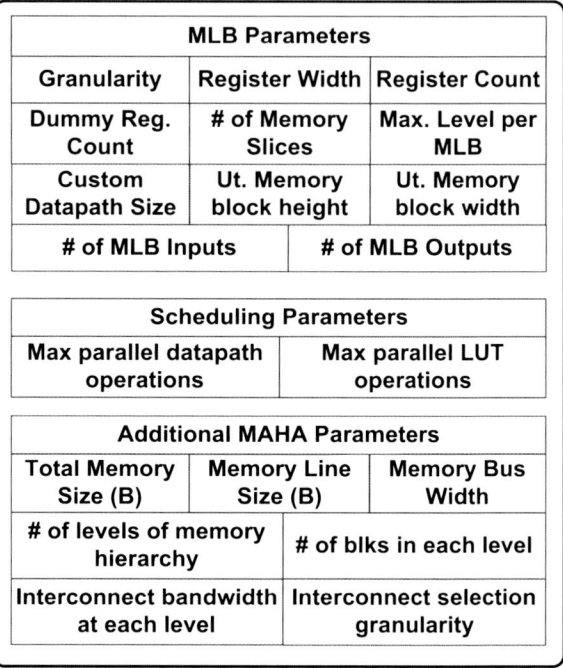

These benchmarks are mostly composed of random logic and offer different degrees of routing complexity to the mapping tool. These benchmarks were mapped to the MBC framework using the MBC software flow for generic reconfigurable framework. The FPGA architecture description file was modified to allow the VPR toolsuite to place and route the *"net"* file that is generated from the MBC software flow. Due to increase in the MLB tile area, it is necessary to accurately estimate the route delay in case of a MBC based reconfigurable platform. However, for accurate estimation of such delay, parameters such as (i) tile area; (ii) switch area and (iii) switch delay need to be determined for the MBC framework. Figure 11.2 shows the breakdown of the MBC tile area and compares the MLB area against the tile area for a baseline 10-cluster 7-LUT FPGA. The MLB tile area was estimated for different combinations of the MLB design parameters (refer to Fig. 11.1), of which only 6 are depicted in Fig. 11.2a. Increase in the MLB tile area demands resizing of the multiplexors and the buffers in unidirectional wires. Sizing was performed for representative resistance and capacitance values for 65 nm interconnect model, considering different tile areas for different MLB configurations (Fig. 11.2b). From Fig. 11.2a and b it can be noted that a MLB configuration with $Max\_LUT\_input = 12$, $Utblkx = 4$, $Slices = 8$, $Row\_subarray = 64$ and $Col\_subarray = 32$ leads to 21 times increase in the tile area along with 10 times increase in switch area and 50% increase in switch delay. Another configuration with $Max\_LUT\_input = 12$, $Utblkx = 2$, $Slices = 8$,

11.2 MLB Architecture Exploration

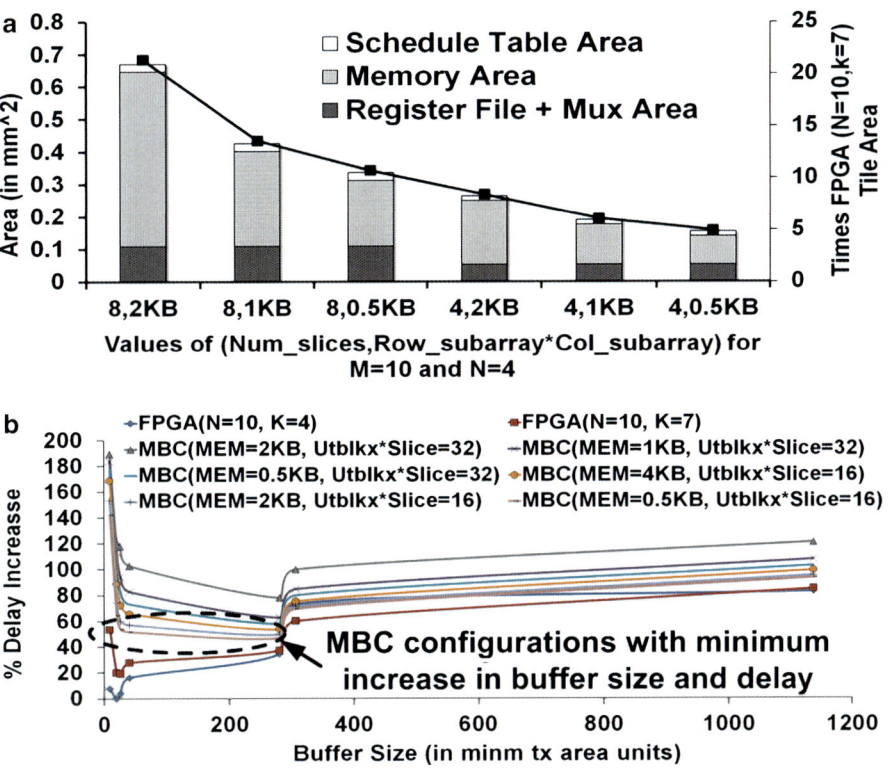

Fig. 11.2 (a) Breakdown of MLB tile area and comparison with FPGA tile area. (b) Estimating the increase in switch area and delay due to higher MLB tile area

Row_subarray = 32 and *Col_subarray* = 32 however, incurs only 8 times increase in tile area and 28% increase in routing delay for the same switch area compared to the baseline FPGA model. Note that for nanoscale technologies, it is beneficial to employ LUTs of larger input size. For the spatial FPGA model, the framework with largest LUT for which models are freely available is for a 7-input LUT. In order to compare between the spatio-temporal MBC model and the fully spatial FPGA at scaled technologies, a 12-input 4-output LUTs in case of MBC and 7-input 1-output LUTs in case of FPGAs have been used.

11.2 MLB Architecture Exploration

This section reports the performance and resources required to map the selected benchmarks to the MBC reconfigurable framework as different design parameters for the framework are varied.

Table 11.1 MBC design space exploration—I

Const: Utblkx=4, Slices=8, Num_DI_subarray=32
Dummy_reg_count=24, MEM=2KB

Var.	Dependent Param.			
# of Partition ips	# of Partitions	Mem Req. (KB)	# of Logic Levels	Max Route Delay (ns)
8	227.76	11.1	4.717	0.3
10	166.67	28.93	3.68	0.34
12	124.12	79.82	3.25	0.28

Table 11.2 MBC design space exploration—II

Const: Max_LUT_input=12, Slices=8, Num_DI_subarray=32
Dummy_reg_count=24, MEM=2KB

Var.	Dependent Param.		
# of Partition ops	# of Partitions	Mem Req. (KB)	Max Route Delay (ns)
1	287.46	40.36	0.37
2	166.9	66.81	0.28
4	124.12	79.82	0.29

11.2.1 Choice of Number of Partition Inputs and Outputs

Following trends were observed (refer to Table 11.1) on varying the partition input size for standard benchmark circuits: (i) As the number of partition inputs is increased maximum number of logic levels in the benchmark circuit is reduced. Total number of partitions also reduces which leads to decrease in the inter-MLB route delay. (ii) On the other hand, total memory requirement increases nearly exponentially. Going from a 8-input partition to a 12-input partition, one would expect a $2^4 = 16$ times increase in the total memory requirement. However, this expected increase is offset by the fact that: (i) there is a linear reduction in the total number of partitions, (ii) not all the partitions have 12 inputs. Simulation results suggest that the combined effect of these two cause only a *8.76X* increase in the memory requirement for standard benchmark circuits when partition input size is increased from 8 to 12 (refer to Table 11.1). Memory requirement increases as 1.21 times for $\sim 2X$ increase in the number of partition outputs. From Table 11.2 it may be noted that across all the benchmarks, the minimum routing delay is achieved for $Utblkx = 2$. The final MLB architecture however supports partitions with output count ($Utblkx$) up to 4 bits wide to minimize the number partitions evaluated and hence improve the energy.

11.2 MLB Architecture Exploration

Table 11.3 MBC design space exploration—III

Const: Max_LUT_input=12, Utblkx=4 Num_DI_subarray=32, Dummy_reg_count=24, MEM=2KB			Const: Max_LUT_input=12, Utblkx=4 Slices=8, MEM=2KB Dummy_reg_count=24		
Var.	Dependent Param.		Var.	Dependent Param.	
# of Mem Banks	# of MLBs	Max Route Delay (ns)	# of MLB Input	# of MLBs	Max Route Delay (ns)
4	24.15	0.34	24	24.15	0.32
6	19.06	0.29	32	17.18	0.28
8	17.18	0.28	40	14.74	0.35
10	16.93	0.33	48	13.2	0.36

11.2.2 Choice of Number of Slices in Function Table

Number of slices in the function table (denoted by *Slices*) dictate the maximum number of partitions that can be evaluated in parallel for a given cycle. With an increase in the value of *Slices*, the total number of MLBs required to map a given design is reduced. The minimum routing delay is however achieved for *Slices* = 8 (refer to Table 11.3). Thus in the final MLB architecture the function table is organized into 8 slices which supports up to 8 parallel partitions in a single execution cycle.

11.2.3 Choice of Number of MLB Inputs and Outputs

Choice of MLB input is guided by the objective of maximizing the number of parallel partitions in the MLBs. From Table 11.3, it can be noted that for small values of *Num_DI_subarray*, the number of MLBs required is more and therefore increases the routing delay. As the number of MLB input is increased, number of MLBs reduces drastically, indicating the increase in the number of parallel partitions in each MLB. If *Num_DI_subarray* is excessively increased, the number of signals that needs to be routed to each MLB is also increased, thus increasing the maximum routing delay for a set of benchmarks. From Table 11.3, it can be noted that the routing delay is minimum for $Num\_DI\_subarray = 32$. As in the case of a clustered FPGA architecture, number of outputs from each MLB in a given cycle is given by "Maximum number of partitions that can be evaluated in parallel×Maximum number of outputs from each partition. For *Slices* = 8 and $Utblkx = 4$, we therefore have $Num\_DO\_subarray = 32$

Table 11.4 MBC design space exploration—IV

Const: Max_LUT_input=12, Utblkx=4, Num_DI_input=32, Slices=8, MEM=2KB			Const: Max_LUT_input=12, Utblkx=4 Num_DI_input=32, Slices=8 Dummy_reg_count=24		
Var.	Dependent Param.		Var.	Dependent Param.	
# of Dummy Reg	# of MLBs	Max Route Delay (ns)	Mem per Bank	# of MLBs	Max Route Delay (ns)
8	18.59	0.31	1KB	18.53	0.32
16	17.93	0.28	2KB	17.18	0.28
24	17.18	0.28	4KB	17.18	0.28
32	16.88	0.29	8KB	17.18	0.28

11.2.4 Choice of the Number of Dummy Registers

Since the partitions are evaluated in a topological manner inside each MLB, it is necessary to use dummy registers to accommodate for signals that cross more than one cycles of execution. As the number of dummy registers (*Dummy_reg_count*) is gradually increased, the total number of MLBs is reduced gradually. For *Dummy_reg_count* greater than 24, the reduction is sufficiently small (refer to Table 11.4). The final MLB architecture therefore supports a maximum of 24 feedforward signals in each cycle.

11.2.5 Choice of Memory Per Slice of the Function Table

With increase in memory size, memory access time increases. Higher memory storage for the function table at each MLB would therefore only lead a negligible increase in the MLB cycle time. Table 11.4 shows the effect of varying the memory per slice (denoted as *MEM*) of the function table. From simulations, it has been observed that only when $MEM < 2\,KB$, there is a 7.2% increase in the total number of MLBs required to map the benchmark circuits. As *MEM* is increased beyond 2 KB, no improvement in MLB count or route delay is observed for $Utblkx = 4$. For $Utblkx = 2$, improvement in routing delay ceases after $MEM = 1\,KB$. The final value of *MEM* is therefore restricted to only 1 KB. From the trade-off analysis provided in Tables 11.1–11.4, the design parameters selected for the final MBC architecture as: (i) *Max_LUT_input*=12, (ii) *Utblkx*=2, (iii) *Num_slices*=8, (iv) *MEM*=1KB, (v) *Dummy_reg_count*=24, (vi) *Num_DI/DO_subarray*=32.

11.3 Comparison with a Fully-Spatial Reconfigurable Architecture

Performance for the MBC based generic reconfigurable framework was validated against a baseline 7-input LUT and 10-LUT clustered FPGA model with 39 inputs. FPGA architectures for 65 nm Predictive Technology Model (PTM) were obtained from [2]. For multi-MLB interconnect framework, F_s (switch-block flexibility) and F_c (connection-block flexibility) was assumed to have the same values as in the baseline clustered FPGA model. The maximum routing delay for a benchmark mapped to the MBC framework determines the inter-MLB cycle time (T_{cycle}). The total time for evaluating a given benchmark is calculated as T_{total}=(# of Logic Levels)$\times T_{cycle}$. Total power expenditure for the MBC framework is estimated as P_{total}=(# of Partitions)$\times P_{partition}$, where $P_{partition}$ denotes the power required to evaluate a single partition. Since the MBC framework evaluates a design over multiple cycles, the total time for evaluating the mapped design is # $ofCycles \times CycleTime$. Energy Delay Product (EDP) for FPGA was compared against that for a MBC framework at supply voltages 1.1 and 0.8 V.

11.3.1 Delay and Power Results

Figure 11.3 shows the break up of the MLB cycle time and the power required to evaluate a single partition. From Fig. 11.3, it can be noted that the MLB cycle times at 65 nm technology node with 1.1 V supply voltage is *680* ps. This includes the placement and routing delay overhead. From Fig. 11.3, the delay through the intra-MLB routing architecture was observed to dominate the cycle time followed by the total memory access delay. The total power required to evaluate a single partition is calculated to be *1.38* mW. This includes power from reading and decoding the schedule table outputs. For estimating the EDP improvement at a lower supply voltage, the cycle times and power requirement at 0.8 V have also been estimated. From Fig. 11.3, it can be noted that the MLB cycle time at 0.8 V increases to *1,132* ps, while the total power requirement for a single partition reduces to *0.89* mW. In order to investigate the technological scalability of the MBC framework, MLB cycle times and power requirements were also estimated using PTM 45 nm [3] technology model file. From Fig. 11.3, it can be noted that at 45 nm, MLB cycle time improves to *560* ps and the power requirement reduces to *0.78* mW. It is important to note here that the delay for the internal routing architecture scales at a higher rate (*30.5%*) from 65 to 45 nm compared to the routing framework used to connect the MLBs or the CLBs as in the case of FPGA.

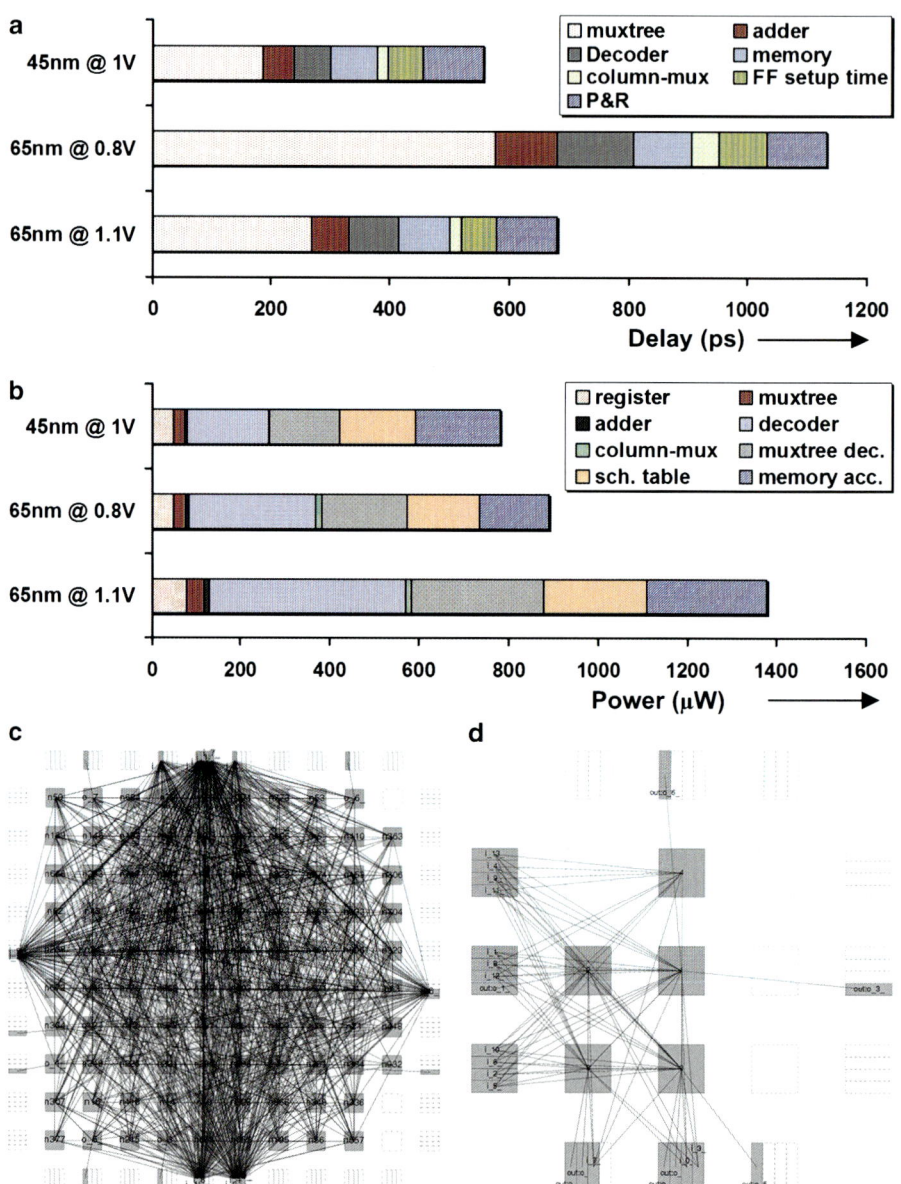

Fig. 11.3 Breakdown of (**a**) delay and (**b**) energy for a single MLB at 65nm technology node. (**c**) alu4 benchmark circuit mapped to (**c**) FPGA and (**d**) MBC frameworks respectively

11.3 Comparison with a Fully-Spatial Reconfigurable Architecture

Fig. 11.4 Scaling of logic and routing delay components across technology nodes for FPGA and MBC frameworks. Starting with the same delay at 65nm, the proposed framework offers 10% more performance benefit compared to FPGA at 45nm

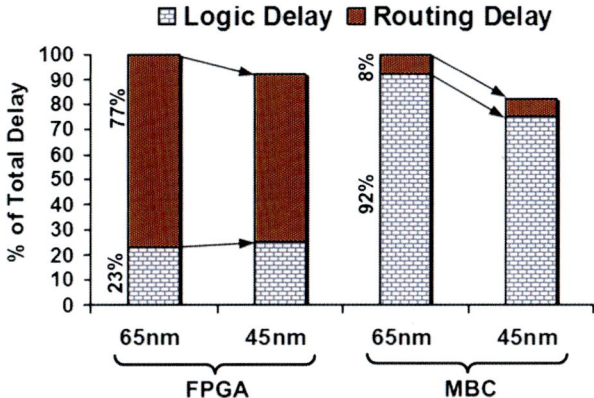

11.3.1.1 Design Overhead for PTM 65nm

Table 11.5 compares the delay, energy and EDP values for mapping the benchmark circuits to the MBC platform as well as to the FPGA framework for two different voltage points. After mapping the benchmark circuits to 12-input LUTs, benchmarks such as *apex4, ex5p* and *bigkey* can be evaluated in a single cycle and there is no overhead for inter-MLB communication. From Table 11.5, it can be noted that the average improvement in delay for all the benchmarks is *56.41%*. From Table 11.5, it can be noted that on an average, the MBC framework requires *84.31%* more energy compared to the FPGA. The EDP estimates hvae also been compared to investigate the energy efficiency of the two frameworks. From Table 11.5, it can be noted that on an average, MBC improves EDP by *29.87%* compared to the FPGA framework at 1.1V. From Table 11.5, at 0.8V, on an average the MBC framework still offers a *34.31%* improvement in performance. The average improvement in energy is *13.65%*. The average improvement in EDP for MBC over FPGA is *51.64%* at 0.8V. Energy efficiency for MBC is more evident at lower supply voltages.

11.3.1.2 Improvement in Resource Usage

From Table 11.5, it may be noted that a migration from the 7-input LUT in FPGA to a 12-input LUT mapping for MBC reduces the total LUT count by *83.72%* on an average. Considerable savings (*75.58%*) is also obtained in the total number of computing elements (FPGA clusters and MLB for MBC framework). Local execution and time-multiplexed use of programmable interconnects in the MBC framework leads to *68.87%* savings in total wirelength compared to the baseline FPGA. This improvement is clearly depicted in Fig. 11.3c,d for a benchmark circuit mapped to both MBC and FPGA frameworks.

Table 11.5 Comparison of MBC and FPGA frameworks for PTM 65 nm

Ckts	% Improv. in # of LUT	% Improv. in # of Clusters	% Improv. in Delay @ 1.1 V	@ 0.8 V	% Incr. in Energy @ 1.1 V	@ 0.8 V	% Improv. in EDP @ 1.1 V	@ 0.8 V
c3540	71.62	71.05	44.38	18.44	232.71	55.88	−85.05	−27.13
c5315	75.89	60.71	14.93	−24.74	–	–	–	–
c6288	86.44	86.30	56.07	41.5	72.84	−19.02	24.07	52.63
c7552	77.34	57.14	33.9	3.08	–	–	–	–
alu4	94.98	95.45	48.85	16.56	−46.57	−74.97	72.67	79.11
apex2	78.30	78.05	33.98	12.08	130.19	7.85	−51.97	5.18
apex4	99.43	98.86	85.28	75.5	−95.29	−97.79	99.31	99.46
des	87.53	80.29	52.44	30.27	−30.86	−67.61	67.12	77.41
ex5p	97.47	96.88	67.77	46.35	59.27	−25.38	48.67	59.97
misex3	89.78	89.89	52.36	22.29	−0.12	−53.21	52.42	63.64
seq	82.46	80.16	30.77	−12.92	89.48	−11.23	−31.17	−0.24
spla	87.92	85.43	53.33	31.56	−26.7	−65.66	65.79	76.5
pdc	90.62	89.84	66.51	45.38	−12.37	−58.95	70.66	77.58
s1423	71.28	63.16	82.37	74.16	336.76	104.63	23.02	47.11
s5378	80.94	64.29	48.8	31.81	222.62	51.15	−65.2	−3.07
s38417	76.80	58.31	76.46	65.49	136.91	11	44.24	61.69
s38584	84.10	48.19	47.17	13.83	–	–	–	–
bigkey	91.19	83.52	88.42	80.72	106.71	−3.16	76.05	81.32
elliptic	66.58	48.65	88.07	80.53	173.44	28.11	67.37	75.06
Avg.	**83.72**	**75.58**	**56.41**	**34.31**	**84.31**	**−13.65**	**29.87**	**51.64**

† Benchmarks mapped with MI=24 to achieve smaller channel width; – Power-aware VPR could not run for these circuits.
* Cutmap reduced the partition count and total energy requirement for these benchmarks.

11.3.1.3 Technological Scalability of Performance

For the case of MBC, only *8.19%* of the total delay is contributed by external routing. Since delay for general-purpose interconnects improves at a lesser pace compared to the logic delay or delay through intra-MLB interconnect, it is expected that the performance improvement obtained through MBC to be more marked for future technology nodes. The FPGA delays at 45nm were estimated using the scaling trend for Altera Stratix family. From 90nm to 65nm, the routing delay scaled by a factor of *0.87*, while the logic delay increased by a factor of *1.11*. The overall delay scaled by a factor of *0.86*. This reduction in the routing delay and increase in the logic delay is due to the change in FPGA logic-block architecture from Stratix II to Stratix III family [4]. Figure 11.4 shows the technology scalability from 65 to 45 nm technology nodes. For benchmark circuits mapped to Stratix FPGA, the average net delay contribution at 65 nm is *77%*, which can be assumed to scale at a value of *0.87* while the logic delay (*23%*) is assumed to scale at a factor of *1.11*. Similarly, the contributions from logic and net delays for MBC at 65 nm are *92%* and *8%* respectively. Since the MLB cycle time improves from 680 ps at 65 nm to 560 ps at 45 nm, the logic delay for the MBC framework scales at a value of *0.82* while the scaling factor for the routing delay is taken to be same as that for FPGA (*0.87*). With these scaling trends, it is worthwhile to note that MBC framework offers *10%* better scalability compared to FPGA platform (Fig. 11.4).

References

1. M. Hansen, H. Yalcin, J. Hayes, "Unveiling the ISCAS-85 benchmarks: A case study in reverse engineering". IEEE Des. Test Comput. (1999)
2. [Online], "iFAR – intelligent FPGA Architecture Repository". http://www.eecg.utoronto.ca/vpr/architectures/
3. [Online], "Predictive Technology Model". http://ptm.asu.edu/
4. [Online], "Stratix Family of FPGAs". altera.com

Chapter 12
Design Space Exploration for MAHA Framework

Abstract Similar to the design space exploration for MBC based generic reconfigurable framework, critical decisions are required to select the design parameters for the MAHA framework. The choice is complicated by the fact, that the design choice should have minimal impact on the normal operation for the memory array which has been instrumented to form the MAHA framework. Moreover, the programmable interconnect resources are more restricted in case of MAHA and choice of interconnect hierarchy affects the success of mapping different applications to this framework. In addition, the size of the custom datapath component present in each MLB in a MAHA architecture needs to be ascertained. Similar to the case of a generic reconfigurable framework, the MBC software flow explores both logic and interconnect components and arrives at a design choice with minimum overhead but with a performance that satisfies the requirement.

12.1 Benchmarks and Simulation Setup

Contrary to benchmarks based on fine-grained random logic used to validate the MBC based stand-alone reconfigurable computing platform, a number of applicationsare selected from different application domains to validate the effectiveness of the MAHA framework. From each of these applications, *critical sections* or *hotspots* which would incur significant power and performance overhead if run on a software platform are mapped to the MAHA framework. These benchmarks are:

i. *Advanced Encryption Standard for 128bits* or *AES-128* and *Secure Hash Algorithm 1* or *SHA-1* from the *Security domain*.
ii. *Motion Estimation*, *Automatic Target Recognition* or *ATR* and *Color Interpolation* from the *Image processing domain*.
iii. *Discrete Wavelet Transform* or *DWT* and *Discrete Cosine Transform* or *DCT* from the signal processing domain.
iv. *Smith Waterman*, *Maximum Entropy Classification* and a synthetic application of *Census* data which belong to the *Informatics* domain.

Table 12.1 Simulation setup for MAHA hardware exploration

Benchmarks		Constant Params		Variable Params	
Domain	Name	Name	Value	Name	Range
Security	AES [1]	Row_subarray	128	Datapath_width	8,16,32
	SHA-1 [2]	Col_subarray	16384	Log_num_slices	0,1,2
Image Proc.	Motion Estimation [3]	Granularity	4	Log_num_levels	2,3,4
	ATR[3]	Reg_count	8	Maxcustom	1,2,4,8,16
	Color interpolation [4]	PR_size	1KB	banks	1,2,4,8
Signal Proc.	2-D DCT[5]	Utblky	64	subbanks	1,2
	DWT[6]	Utblkx	16	mats	1,2
Infomatics	Smith Waterman [7]			subarrays	1,2,4,8
	Max entropy Classification [8]			Buswidth	64,128, 256,512, 1024
	Census				

Details of the simulation setup used for this exploration are provided in Table 12.1. As listed in Table 12.1, it is assumed that the memory size per subarray of the instrumented memory is $Row\_subarray \times Col\_subarray = 256KB$.

12.2 MLB Architecture Exploration

Figure 12.1 shows the methodology for exploring the hardware architecture of the MAHA framework and arriving at the final memory configuration which would support the MAHA framework. As illustrated in Fig. 12.1, the final design choice is guided by the optimization goal, i.e. minimizing the area, energy or maximizing the performance of the mapped applications. Through synthesis and SPICE level simulations at PTM 45nm [9] technology node, the area is first found out, energy and performance for the individual components of the MLBs and the routing switches which would be used to route the inter-MLB signals for each MLB and routing switch configuration. Then during exploration step of the software flow, each configuration, these area, performance and energy numbers are collated to obtain the total area overhead, energy requirement and performance for the mapped application. Finally based on the optimization objective, one of the configurations is chosen for the final implementation.

12.2.1 Choice of the Datapath Width

Different applications have different requirements for the width of the custom datapath. For example, *SHA-1* benefits from a 32-bit datapath while *DCT* requires a 16-bit datapath. Choosing a datapath of appropriate size is therefore a trade-off

12.2 MLB Architecture Exploration

Fig. 12.1 Methodology for exploring the hardware design space of the MAHA framework

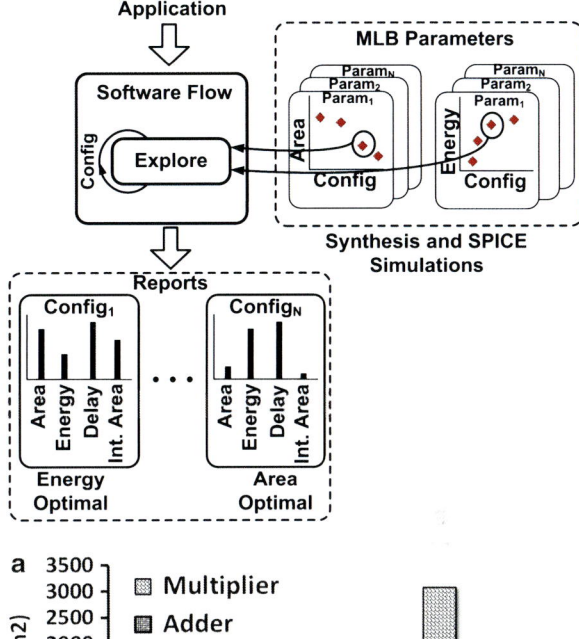

Fig. 12.2 (a) Area and (b) Energy overhead for different components of the custom datapath for different datapath widths

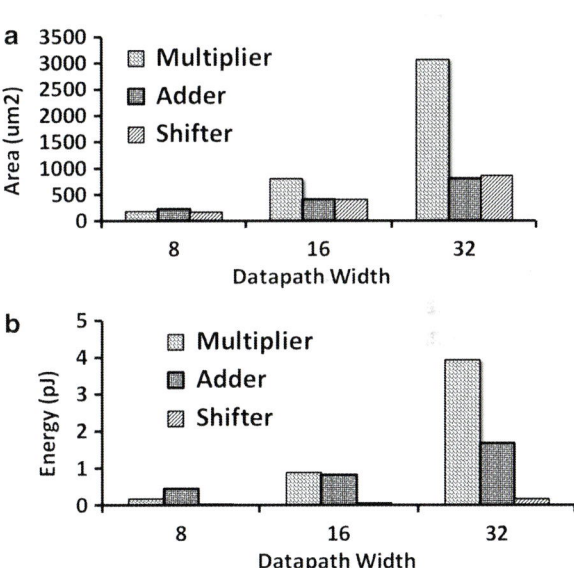

between performance and design overhead. The width for the datapath is varied as illustrated in Fig. 8.3 as 8, 16 and 32 and noted the corresponding impact at 45nm technology node. The results are provided in Fig. 12.2.

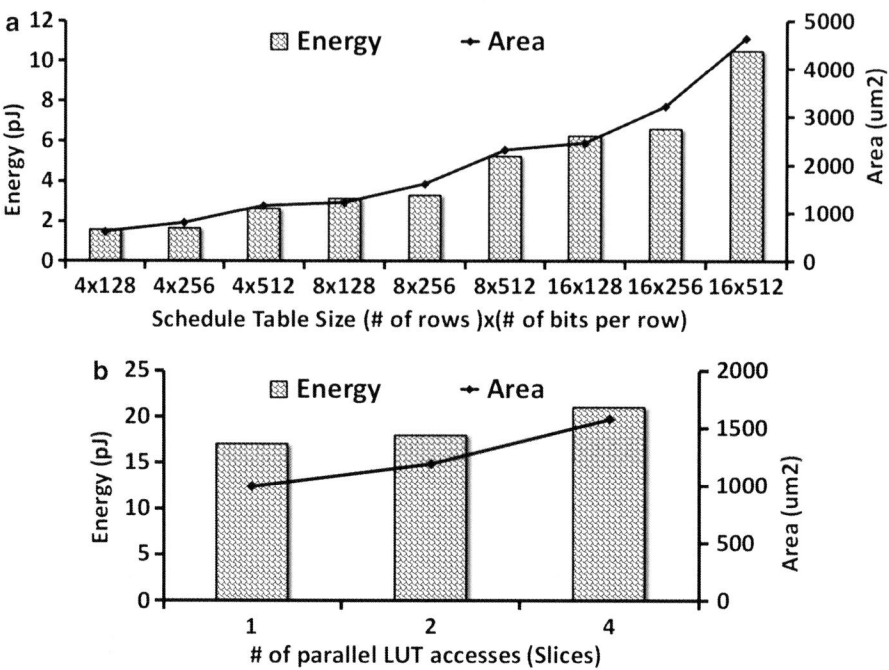

Fig. 12.3 (a) Area and (b) Energy overhead for different sizes of the *schedule table* and *muxtree*

12.2.2 Choice of Number of Levels of Execution Inside MLB

Increasing the number of sequential operations inside the MLB increases the temporality inside the MLB, at the cost of increasing the *schedule table* mpact on the *schedule table* area and energy.

12.2.3 Choice of Number of Slices in Function Table

Number of parallel LUT accesses from the *function table* are denoted by the parameter *Slices*. Increasing this design parameter, increases the number of parallel operations inside the MLB at the cost of increase in hardware overhead. It directly increases: (i) the *schedule table* area since each row of the *schedule table* has to store more instruction opcodes and (ii) the *muxtree* area which has to select more operands for the parallel operations. The increase in area and energy for the *schedule table* and *muxtree* for different values of *levels* and *slices* in the MLB design are shown in Fig. 12.3a and b respectively.

12.2.4 Choice of the Memory Hierarchy and Interconnect Bandwidth

Memory arrays are conventionally organized for maximum integration density and/or minimum energy dissipation. It is however important to explore the optimal memory organization for reconfigurable computing. For example, while some applications might require more local communication among its sub-operations, other might demand more global communication. For the first case, the number of *subarrays* and the bandwidth at each *subarray* must be high. The second situation however demands more number of *banks* and large *Buswidth* value. The memory organization and the *Buswidth* thus directly impacts the routability for the mapped application. By varying the parameters that define the memory hierarchy, such as number of banks, subbanks, mats and subarrays and the *Buswidth*, it was noted that as the *Buswidth* is increased, the routability improves. Moreover, for the same number of physical resources (i.e. *Banks×Subarrays per bank*) it was more beneficial to provide more number of *Banks* than *Subarrays per bank*.

12.2.5 Variation in Energy-Efficiency with Varying MLB Design Parameters

MLB parameters such as datapath width, number of levels and number of slices have considerable effect on (i) area for each MLB; (ii) number of MLBs; (iii) total MLB energy; (iv) total latency per input vector; (v) total interconnect area and (vi) total interconnect energy. While the actual impact on area and energy depends on the area and energy for the *function table* access and therefore the exact memory technology used, it is possible to note the impact on the number of MLBs and performance in cycles for the mapped applications as the design parameters are changed. Figure 12.4a and b shows the effect on latency (in terms of cycles) and MLB count for benchmark *SHA-1* as the MLB design parameters are varied. For a particular set of design parameter values; the minimum latency and MLB count points have been plotted. Missing point at a particular configuration indicates such a configuration did not go through *clustering* or *routing* step due to insufficient logic or routing resources. As expected, *slices=2* has lower latency and lower MLB count compared to *slice=1* and *slice=0*, since increasing the number of slices, increases the number of parallel operations per clock cycle. A similar effect going from *datapath=16* to *datapath=32*. As the number of levels is increased, the MLB count decreases, since each MLB is able to accommodated more operations. Finally, latency should not ideally change with the number of levels. However, for *slice=1*, it may be noted increase in latency due to increase in *levels*. The reason being, the for *level=4*, the number of vertices in each MLB increases which increases the requirement for MLB I/O. Hence, a *slice=1,level=4* can only pass routing

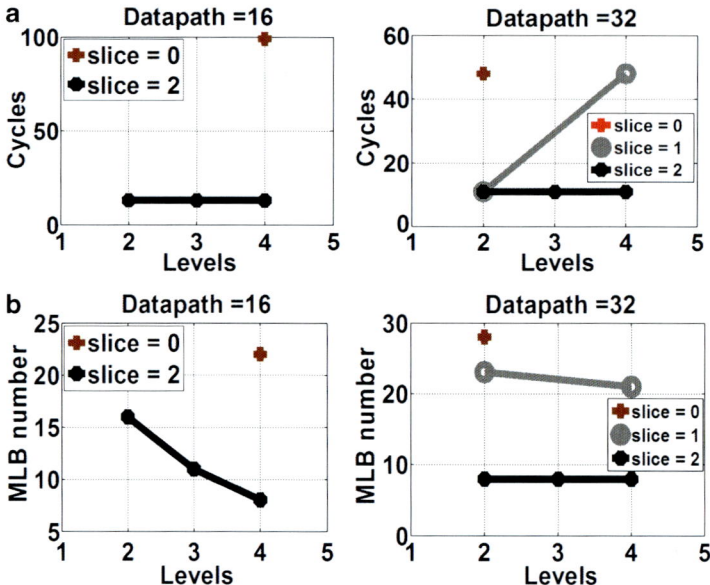

Fig. 12.4 Variation in (**a**) latency and (**b**) MLB count for *SHA-1* with variation in (i) levels; (ii) slices and (iii) datapath width

if the number of parallel operations is restricted, i.e. *Maxcustom* is less which negativelyimpacts the latency.

References

1. T. Good, M. Benaissa, "692-nW advanced encryption standard (AES) on a 0.13 μm CMOS". IEEE Trans. Very Large Scale Integrat. Syst. (2009)
2. [Online], "Secure Hash Algorithm - 1". http://en.wikipedia.org/wiki/SHA-1
3. H. Singh, M. Lee, G. Lu, F.J. Kurdahi, N. Bagherzadeh, E.M. Chaves Filho, "MorphoSys: anintegrated reconfigurable system for data-parallel and computation-intensive applications". IEEE Trans. Comput. **49**(5), 465–481 (2000)
4. [Online], http://vigir.missouri.edu/~gdesouza/Research/ColorCCD/ColorInterpolation.pdf
5. U. Meyer-Baese, "Digital Signal Processing with Field Programmable Gate Arrays" (Springer, Heidelberg, 2007)
6. K. Andra, C. Chakrabarti, T. Acharya, "A VLSI architecture for lifting-based forward and inverse wavelet transform". IEEE Trans. Signal Process., 966–977 (2002)
7. [Online], "Smith-Waterman Algorithm". http://en.wikipedia.org/wiki/Smith--Waterman_algorithm
8. K. Nigam, J. Lafferty, A. McCallum, "Using Max Entropy for Text Classification", in *Intl. Joint Conf. on Artifical Intelligence*, 1999
9. [Online], "Predictive Technology Model". http://ptm.asu.edu/

Chapter 13
Preferential Memory Design for MBC Frameworks

Abstract Since memory is primarily used for computation in a memory based reconfigurable framework, memory access is dominated by read rather than write, which occurs only during the process of application mapping. For a given technology, optimizing a memory cell for both read and write pose contradictory requirements (Mukhopadhyay et al., IEEE Trans Comput Aided Des Integrated Circ Syst 24(12), 2005). However, exploiting the read-dominated memory access pattern in reconfigurable frameworks, it is possible to come up with a novel memory cell design which offers better read power and performance at the cost of increased write power and performance. Dense memory arrays with such a memory cell would therefore offer better energy efficiency. Details of these circuit-level optimizations corresponding to CMOS based SRAM technology and emerging STTRAM technology are provided in this chapter. This chapter also presents circuit techniques which skew the memory cells be more energy-efficient when storing logic "0" rather than logic "1". Application mapping heuristics are presented which can leverage this asymmetric memory behavior for further improving the energy-efficiency of the applications mapped to the MBC framework.

13.1 Case Study—I: SRAM Array

This section describes the preferential design approach for CMOS based SRAM. Conventionally SRAMs are optimized for both read and write. However, the read-dominant usage opens the opportunity to optimize these memories primarily for read, improving read power, performance and reliability. The cost is a degradation in write power, performance and reliability which can be easily addressed through a slightly increased write time.

In order to improve the read performance and minimize the read power, a 6-T SRAM cell (Fig. 13.1a) which employs one-sided write and dynamic read methodology is proposed [1]. This improves the EDP during normal operation of the reconfigurable framework at the cost of higher write power during infrequent

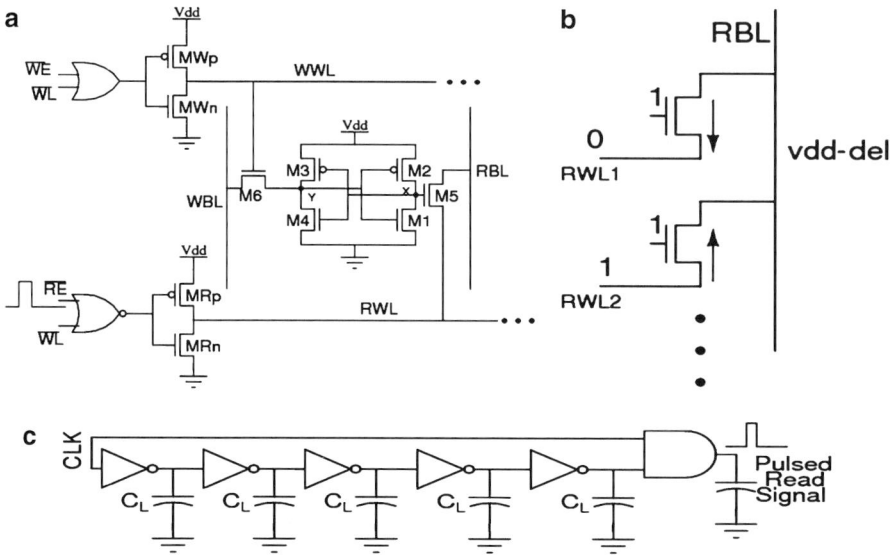

Fig. 13.1 (a) The proposed 6-T SRAM cell; (b) Pathological case showing the flow of static current from an unselected cell to the selected cell; (c) Pulse generation circuit for read operation

reconfigurations of the platform. The SRAM cell benefits the MBC framework in terms of increased SNM, more tolerance to process variability and reduced read power consumption. Also the proposed cell being essentially a 6-T structure the area overhead associated is less than in 7T or 8T structures proposed earlier [2]. However, writing to the cell is single-ended and requires word line boost up methods thereby consuming more write power than the conventional 6-T structure. So the proposed structure's energy efficiency over the conventional case is a direct function of the read write ratio. The proposed cell considers reading by controlling the source of the access transistor for read. This concept has been explored earlier in [3]. However, the solution proposed in [3] suffers from the fact that the bitline for the cells are coupled via the high impedance source. This might lead to corrupting data of the unselected cells and giving false reads. This problem is eliminated by maintaining the source of the access transistor at low impedance and using a pulsed read mechanism. This decouples the cells effectively. More details on the read/write mechanism of the proposed SRAM cell are provided in [1].

13.1.1 Read-Stability

The proposed memory structure has an improved read-stability. This is due to decoupling of the read and writes such that the cell being read from is isolated from the bitline so that there is no issue of the cell flipping while reading. The SNM

13.1 Case Study—I: SRAM Array

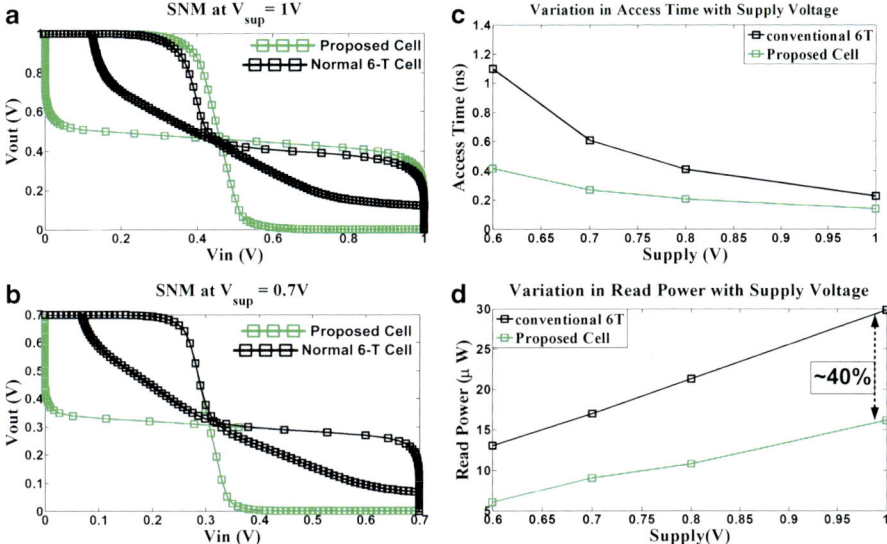

Fig. 13.2 (a) The proposed cell offers a higher SNM compared to the conventional 6-T SRAM at nominal voltage; (b) The improvement in SNM becomes more marked at lower supply voltages; (c) (a) The proposed memory cell has improved access time compared to the conventional 6-T SRAM cell. The improvement in access time is higher for lower supply voltages; (d) The proposed cell also achieves a 40% reduction in the read power at nominal supply voltage

curves in Fig. 13.2a illustrate the read-stability advantage of the cell at supply = 1 V. At lower supplies, the distinction is even more prominent as shown in Fig. 13.2b. This is because of the fact that due to the smaller supply voltages, read disturb will dominate even more for standard 6-T cell whereas the charge at the node "X" still remains decoupled for proposed cell structure.

13.1.2 Read Energy as a Function of the Pulsed Read

In order to address the pathological case as illustrated in Fig. 13.1b and minimize the static power dissipation for the worst case condition, a pulsed read mechanism is proposed. Details of the pulsed read mechanism can be obtained from [1]. The amount of static power dissipated in the proposed memory cell is a direct function of the length of the pulse. For reliable operation of the proposed cell and to achieve power savings over a conventional 6-T cell, a pulse width of 150 ps was selected.

13.1.3 Operation at Higher Frequency

Figure 13.2c,d indicate that the proposed cell is clearly a better choice for low voltage and high frequency operations. Even considering 200 mV of necessary drop instead of the customary 100 mV drop for the differential detection across SRAM, the cell has 2X lower access time than the 6-T structure. This essentially means an increased frequency of operation is possible for the proposed cell. More details on high frequency of operation of the proposed SRAM cell can be obtained from [1].

13.1.4 Writability

Writing to this cell being single ended, to ensure reasonable writing time, the write bit line and the word line are overdriven. The cell voltage is maintained at 0.6 V whereas the W_{BL} and W_L are maintained at 1 V while writing. Whereas the unselected cells of the selected column do not have problems for such a scheme this can give rise to stored data being flipped for the unselected cells along selected row. This may be avoided by maintaining the unselected cells along the selected row at higher voltage ($V_{dd} = 1$ V). More details on the writability of the proposed cell can be obtained in [1].

13.1.5 Asymmetric Behavior

The proposed memory cell is asymmetric with respect to the power consumed during read "1" and read "0" operations. Due to higher read power during the read "1" operation, it is intended that the memory location being accessed contains logic "0" rather than logic "1". Furthermore, for the read "1" scenario, it is desired that the unselected rows in the same column have higher number of logic "0" to reduce the static power dissipation during the pulsed read operation. Figure 13.3a shows the reduction in read-"1" power as the number of cells storing "0" in the same column increases. This asymmetric behavior of the memory cell in storing "0"s and "1"s is exploited by the content-aware mapping heuristic proposed earlier to skew the LUTs contents for improving the read energy.

13.1.6 Improvement in EDP for MBC and EMB Based Heterogenous FPGAs

For the proposed MBC based stand-alone reconfigurable framework, it may be noted that for conventional 6-T SRAM based MBC framework, the memory array

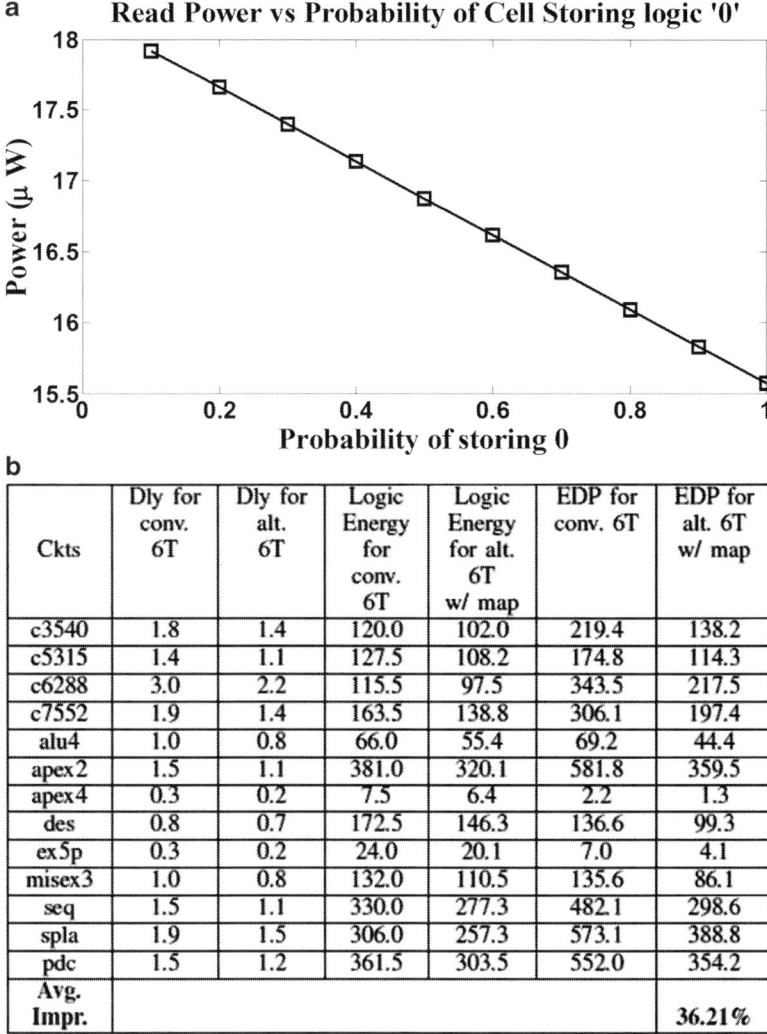

Fig. 13.3 Reading from a cell storing logic "0" requires less read power compared to a cell storing logic "1". The total read power thus decreases as the probability of the memory locations storing "0" increases

contributes *29.5%* to the MLB cycle time and *24.3%* to the MLB power per cycle. With the preferentially designed SRAM cell, it may be noted that the MLB cycle time improves by *12.7%* and the power improves by *5.2%* respectively. Since the proposed SRAM cell improves the performance and power for each memory access, an EMB based heterogenous FPGA [4, 5] can greatly benefit from the use of the proposed memory design. Standard benchmark circuits were mapped to 12-input LUTs using the *Heteromap* mapping algorithm proposed in [6]. Figure 13.3b shows the delay and energy results for the combinational MCNC benchmarks mapped to the EMB based heterogenous FPGA framework. From Fig. 13.3b, it may be noted that the read-optimized memory cell along with content-aware mapping approach achieves a *36.21%* improvement in EDP in a EMB based heterogeneous FPGA framework. Since the routing energy remains the same for the two scenarios, only the logic energy with and without the proposed co-design approach has been compared.

13.2 Case Study—II: STTRAM Array

STTRAM [7–9] is a promising non-volatile memory technology which has a number of distinct advantages over the other prevalent memory technologies. Having essentially a 1T-1R structure it forms a dense array with high integration density and being magnetic in nature is tolerant to particle hits. In these features it is better than the SRAM. Being non volatile, it does not require refreshing like DRAM. It has a high write endurance, which makes it a better candidate than Flash memories. Another feature of STTRAM is its scalability which makes it an attractive option at scaled dimensions. In the nanometer nodes, leakage is one of the primary forms of energy dissipation. The STTRAM structure has zero standby leakage. All these attributes make STTRAM extremely attractive as a reconfigurable computing fabric. It has been demonstrated that a spatio-temporal MBC architecture realized with 2-D STTRAM arrays, a STTRAM memory cell optimized for read and a content-aware application mapping approach can be integrated into an unified design flow to realize an energy efficient reconfigurable framework [9]. In the proposed non-volatile MBC framework, the function table inside each MLB is realized using STTRAM arrays. This offers the following benefits: (i) since the function table holds the configuration for the partitions, it occupies the maximum area inside a MLB. A small footprint of the Magnetic Tunneling Junction (MTJ) device [7] ensures that the area occupied by this memory array is minimized; (ii) non-volatile nature of the STTRAM array ensures that configuration bits stored in the function table are retained when power is turned down; (iii) high read performance and low read power for the STTRAM array results in considerable EDP improvement for a STTRAM based non-volatile MBC framework.

13.2.1 Novel STTRAM Cell Design

Exploiting the read dominated nature of the memory access in the MBC framework, the STTRAM cell can too be optimized for read operation at the cost of write. Similar to the 6-T SRAM cell, the design space for STTRAM is constrained by the readability and writability conditions i.e. tunnel magnetoresistance (TMR) ratio and write current requirement. Given a MTJ, the choice of the access MOSFET width

13.2 Case Study—II: STTRAM Array

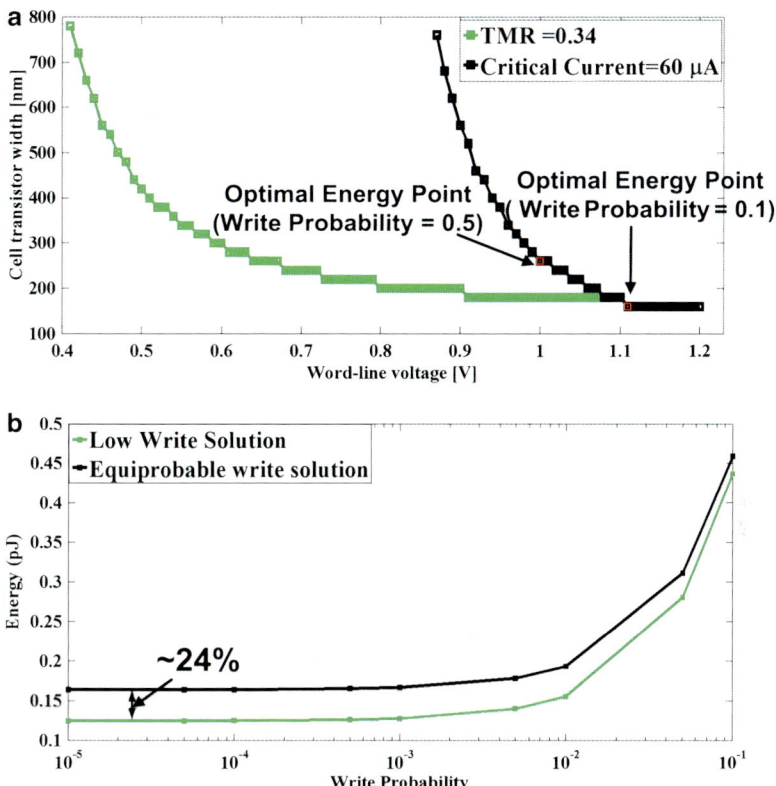

Fig. 13.4 (a) Design of STTRAM cell for MBC framework to achieve optimal read energy; (b) read energy with varying write probability

(W) and its wordline voltage (V_{WL}) can be used to navigate the design space of TMR and write-current. A key aspect of the solution is its dependence on the read–write probability. Figure 13.4a shows two solutions corresponding to write probabilities of 0.5 and 0.1, respectively. A larger write probability means a solution with larger width and smaller V_{WL} as write has a quadratic dependence on V_{WL}. From read perspective, a solution with lower width is preferred due to less leakage power dissipation. For MBC with read-dominant access pattern, the $W - V_{WL}$ configuration corresponding to equiprobable condition is not an optimal choice as it dissipates higher read energy (Fig. 13.4b). Hence the optimal energy point selected is one corresponding to low write probabilities which provides much lower read and total energy at the expense of increased write energy. Figure 13.4b shows 24% saving in total energy for write probability of 10^{-5} for an 8-bit 64×64 memory array with a read access time of 400 ps.

13.2.2 Improvement in EDP for STTRAM Based MBC

Simulations with MTJ have been performed at 65 nm node with Resistance-Area product $30\,\Omega - \mu m^2$. The sizes of the MTJ devices have been taken as $50 \times 90\,nm^2$ which requires approximately $60\,\mu A$ of switching current assuming current density of $10^6\,A/cm^2$. The high and low resistance states are represented by $11.1\,k\Omega$ and $6.67\,k\Omega$, respectively. For the required STTRAM characteristics, the MTJ device characteristics as reported in the paper are used. The resistive values for the parallel and anti-parallel states at the read and write voltages are of consequence to us. The abstracted resistive behavior of the MTJ is simulated in conjunction with NMOS device at 65 nm predictive technology model. For all measurements, the simulation for the STTRAM cell is done using HSPICE. The transistor was designed so as to be able to drive the switching current under both write "0" and write "1" conditions. To obtain the solution for varying write probabilities, first a host of simulations with the high and low resistance is performed for a range of V_{WL} and W. The solution space has to be extracted from the generated design space considering the constraints on minimum TMR and switching current. This work considers the minimum TMR and switching current requirements of 0.34 and $60\,\mu A$, respectively. Corresponding to this extracted feasible design space of $V_{WL} - W$, the read, write, active leakage and total energy of STTRAM array are evaluated. The energy evaluation considers the read and write probability ratios. The $V_{WL} - W$ combination which gives the minimum energy is identified as the optimal energy solution. For demonstration of the results, write probabilities of 0.5 and 0.1 are selected to evaluate the design points. The read and write energies for "0" and "1" are computed for these design points.

Delay and energy requirement for the CMOS elements of the MLB were obtained through SPICE simulations using predictive models at 65 nm technology node [10]. Delay and energy estimates for the MBC framework with the read optimized STTRAM cell was obtained from the integrated automation flow described in Chap. 10. These were then compared against that for a 7-input LUT, 10-LUT cluster FPGA model at 65 nm technology node [11]. Figure 13.5a,b shows the improvement in performance and energy-delay product for STTRAM based MBC over the baseline CMOS FPGA model. As may be noted from Fig. 13.5a, for standard benchmark circuits on an average the MBC framework improves the performance by *45.4%*. Figure 13.5b compares the EDP values between the two frameworks. The non-volatile MBC framework achieves a *5%* improvement in EDP over the CMOS FPGA framework. The performance and EDP computation includes the cell optimization for read operation. The EDP improvement is further enhanced through the content-aware mapping step which skews the LUT contents to have more logic "1" than logic "0"s. As a result of this content-aware mapping, the average EDP improvement was calculated to increase from 5% to *6.64%*.

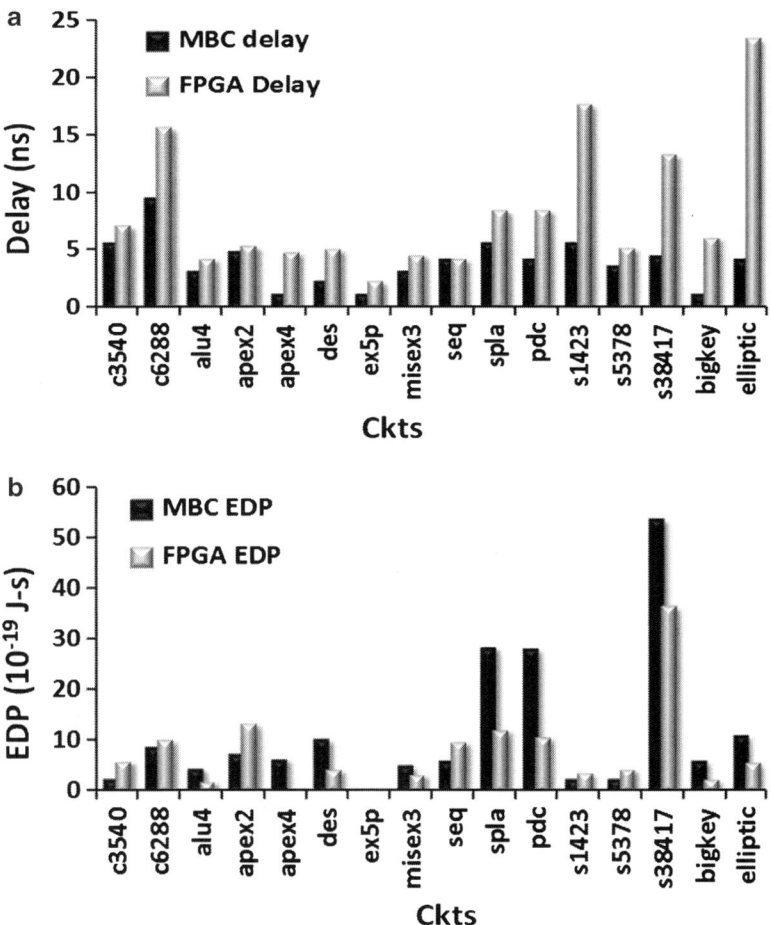

Fig. 13.5 Improvement in: (**a**) delay and (**b**) Energy Delay Product (EDP) for STTRAM MBC over conventional SRAM-based FPGA

13.3 Optimization in Application Mapping

Leveraging the asymmetric behavior exhibited by the SRAM and STTRAM memories presented earlier, this section describes the software opportunities for improving the energy requirement of the MBC framework. The main idea is to modify the contents of the read-only memory during application mapping such that they contain more logic "0" instead of logic "1". This allows the memory read operation to be more energy-efficient and reliable.

Algorithm 4 Content–aware Mapping to maximize the percentage of logic "0" in the LUTs
---
1: **Input:** 1) DAG representation G(V,E) of the partitioned netlist. 2) Truthtable T_j for the individual partitions ($Part_j$).
2: **Objective:** To maximize the percentage of logic "0" in the truthtables of the individual partitions.
3: Levelize partitions. Let the total number of levels be L.
4: **for** $i = 1$ to L **do**
5:    **for** $Part_j \in Level_i$ **do**
6:       **if** Output from $Part_j \notin$ PO or Flop Inputs **then**
7:          **if** in T_j, # of ones > # of zeros **then**
8:             Flip 0's and 1's.
9:             Rearrange the 0 and 1 locations for the LUTs at the fanout of $Part_j$
10:          **end if**
11:       **end if**
12:    **end for**
13: **end for.**

13.3.1 Content-Aware Mapping

The main computing fabric for the MBC framework is memory. For memory technologies such as SRAM and STTRAM, it is possible to skew the memory cells such that memory cells containing 0/1 have better read energy compared to cells storing 1/0 [1, 9]. In such scenarios, it is possible to intentionally skew the LUTs stored in the 2-D memory array (function table) to contain more 0/1 than the other. This leads to an improvement in the average memory access energy. A content-aware application mapping methodology can therefore exploit this asymmetric nature of memory cells and amplify the energy savings by storing more logic "0" than logic "1" or vice versa. Note that the concept of skewing LUT content has been explored for diverse reasons in the context of FPGAs. In [12], the authors attempt to improve the soft-error tolerance of FPGA configuration bits by storing more logic zeros in the LUTs. In [13], the authors have proposed to reduce the leakage power consumption in FPGA by correct selection of polarity for the configuration bits. A greedy heuristic can thus reduce the active power consumption in the MBC framework by preferentially skewing the ratio of logic "0"s to logic "1"s in the LUTs.

13.3.2 Heuristic for Content-Aware Mapping

Algorithm 4 presents the pseudocode for maximizing the percentage of logic "0"s stored in the LUTs mapped to the memory array. The procedure to maximize the percentage of logic "1"s follows similarly. Input to Algorithm 4 is the hypergraph representation (G(V,E)) of the partitioned netlist and the truthtables for the

13.3 Optimization in Application Mapping

Table 13.1 Improvement in stored logic "0" count achieved through skewing of LUTs

Ckts	# of Partitions	% of Logic "0"s Before Opt	% of Logic "0"s After Opt	%Impr
c3540	80	44.44	73.25	28.81
c5315	85	35.17	75.04	39.87
c6288	77	17.71	82.87	65.16
c7552	109	28.45	74.68	46.23
alu4	44	27.45	91.98	64.53
apex2	254	33.87	90.24	56.37
apex4	5	46.68	64.8	18.12
des	115	24.02	75.99	51.97
ex5p	16	46.51	96.51	50
misex3	88	21.78	94.99	73.21
seq	220	27.24	90.06	62.82
spla	204	25.5	88.83	63.33
pdc	241	27.36	91.24	63.88
s1423	41	54.21	68.3	14.09
s5378	106	42.76	82.01	39.25
s38417	794	31.22	77.33	46.11
s38584	745	31.36	81.42	50.06
bigkey	160	18.61	81.55	62.94
elliptic	102	42.92	81.55	38.63
Avg.				**49.23**

individual partitions. The partitions are first levelized following their topological order in the netlist. Then beginning from partitions present in the first level, the truthtable of the individual partitions are examined. If the truthtable is found to contain more logic "1"s, then each LUT location is inverted. This is only done for LUTs which do not drive any primary output of the circuit or input to any internal state element. This is done to avoid reversing the polarity of the primary or state element output. In order preserve the correctness of the logic in the following levels, truthtable for the LUTs in the fanout of the modified LUT are rearranged. This rearrangement involves a simple reordering of the 0 and 1 locations inside the truthtable and does not modify the total 0 or 1 count.

13.3.3 Improvement in Logic "0" Count

The effectiveness of the mapping approach has been validated using a set of standard circuits chosen from the ISCAS'85, ISCAS'89 and MCNC benchmark suites mapped to 12-input 4-output LUTs. Table 13.1 shows the percentage of logic "0" count in the LUTs for the target circuits before and after content-aware mapping. From Table 13.1, it may be noted that the mapping heuristic achieves almost *49%* increase in the percentage of logic "0" count stored in the LUTs.

References

1. S. Paul, S. Chatterjee, S. Mukhopadhyay, S. Bhunia, "A Circuit-Software Co-design Approach for Improving EDP in Reconfigurable Frameworks", in *IEEE/ACM Intl. Conference on Computer-Aided Design (ICCAD)*, 109–112 (2009)
2. L. Chang et al., "Stable SRAM Cell Design for the 32nm Node and Beyond", in *Symp. on VLSI Technology*, 2005
3. J. Singh, J. Mathew, S.P. Mohanty, D.K. Pradhan, "Single Ended Static Random Access Memory for Low-Vdd High-Speed Embedded Systems", in *VLSI Design*, 2009
4. J. Cong, S. Xu, "Technology Mapping for FPGAs with Embedded Memory Blocks", in *Intl. Symp. on FPGAs*, 1998
5. S.J.E. Wilton, "SMAP: Heterogeneous Technology Mapping for Area Reduction in FPGAs with Embedded Memory Arrays", in *Intl. Symp. on FPGAs*, 1998
6. J. Cong, S. Xu, "Performance-driven technology mapping for heterogeneous FPGAs". IEEE Trans. Comput. Aided Des. Integrat. Circ. Syst. **19**(11), 1268–1281 (2000)
7. T.W. Andre et al., "A 4-Mb 0.18-m 1T1MTJ Toggle MRAM with Balanced Three Input Sensing Scheme and Locally Mirrored Unidirectional Write Drivers", in *Intl. Solid-State Circuits Conference*, 2004
8. S. Paul, S. Bhunia, "Computing with Nanoscale Memory: Model and Architecture", in *Intl. Symp on Nanoscale Architecture*, 2011
9. S. Paul, S. Chatterjee, S. Mukhopadhyay, S. Bhunia, "Nanoscale Reconfigurable Computing Using Non-Volatile 2-D STTRAM Array", in *Intl. Conf. on Nanotechnology*, 2009
10. [Online], "Predictive Technology Model". http://ptm.asu.edu/
11. [Online], "iFAR – intelligent FPGA Architecture Repository" http://www.eecg.utoronto.ca/vpr/architectures/.
12. S. Srinivasan, A. Gayasen, N. Vijaykrishnan, M. Kandemir, Y. Xie, M.J. Irwin, "Improving Soft-Error Tolerance of FPGA Configuration Bits", in *ICCAD*, 2004
13. J.H. Anderson, F. Najm, "Active leakage power optimization for FPGAs". IEEE Trans. Comput. Aided Des. Integrat. Circ. Syst. **25**(3), (2006)

Chapter 14
PLA Based Application Mapping in MBC

Abstract This chapter presents a unique implementation of MBC framework which realizes function not by LUTs but rather by programmable logic arrays (PLA). While LUTs sizes tend to be exponentially large with large number of inputs, PLA sizes increase at a much lower rate, thereby making them attractive for representing functions with large number of inputs. From the implementation perspective, the benefit is exponentially smaller memory size compared to a LUT based approach. This leads to considerable improvement in performance and energy for the MBC framework. The challenge is however, conventional random access memories cannot be used to store and retrieve the PLA representation. Content-addressable memories (CAM) are ideal candidates for storing the PLA representation. This chapter describes the CAM based MBC architecture and the corresponding software flow.

14.1 Function Representation as PLA

Although the integration density for present and emerging memory technologies are expected to significantly improve in the future [1], a judicious use of the memory available is always welcome since it (a) reduces the total memory requirement, which leads to faster cycle time, (b) improved energy/vector requirement and (c) potential to map larger application at each computational unit of a memory-based computing framework. Although most functions can be efficiently represented using a PLA-based representation, a reconfigurable platform using such a representation requires a different memory structure such as Content Addressable Memory (CAM), which has built-in search capabilities. Modern CAM implementations are usually large with word length ranging from 36 to 144 bits and address space from 7 to 15 bits [2]. In spite of large size, such CAM designs allow access times as low as 0.25 ns [3]. Since a CAM-based PLA representation of a function can lead to better utilization of the storage space, it is worthwhile to investigate the effectiveness of CAM structure for memory based computing.

Although, efforts [4] have been made to configure a small storage at each CLB (in a conventional SRAM-based FPGA) both as a CAM and a LUT for efficient functional representation, the limited number of inputs to a CLB fail to exploit the high integration density offered by modern CAM designs. For efficient utilization of the die real estate, it is therefore desirable to have a large content addressable storage, which can be configured as multi input-output PLAs. Embedded System Block (ESB) of the APEX20K from Altera Corporation incorporates such an embedded memory, which can perform as a CAM-based CPLD (Complex Programmable Logic Devices) structure [5]. However, such an implementation maps only the onset or offset of a function and cannot exploit the optimization obtained by consideration of don't care terms. A novel implementation of the memory based computing paradigm can therefore leverage the high integration density of modern CAMs. The idea is to realize the *function table* inside the MLB using a dense 2-D CAM array and store the optimized PLA representation of the partitions in the CAM array. Similar to LUT based MBC framework, these partitions are evaluated in a topological order over multiple clock cycles. A hybrid CAM-LUT based function representation has also been investigated which further reduces the memory area by selectively storing some partitions in the CAM, while others in LUT.

14.2 Function Representation Using CAM Based MBC Framework

14.2.1 *Optimized PLA Using Ternary CAM (T-CAM)*

In order to evaluate a partition in memory given its PLA representation, it is necessary to match the inputs to the PLA representation that is already stored in the function table. One hurdle for such an optimized implementation is that for many entries in the ON-set of a function, one or more function inputs may have "X" (don't care) values. This challenge is addressed by modifying the conventional CAM structure to store and match "X" values regardless of the input bit, similar to [2]. Such a CAM structure is often referred to as Ternary CAM (T-CAM). However, in order to store 3 logic values (0, 1 and X), each memory cell must contain two storage bits. Thus, the data stored in a T-CAM cell can be matched against either 1, 0 or X on the input lines. Figure 14.1 shows the NOR-based implementation of a single T-CAM cell with the corresponding encoding for the logic levels. Based on the observation that a collection of T-CAM cells have a structure similar to the "AND" plane of a PLA, it is possible to construct a PLA structure in the memory with the "OR" plane being constructed from the normal SRAM cells, which hold appropriate values for the output response.

a **Ternary Encoding for NOR CELL**

Value	Stored		Search Bit	
	D	D'	SL	SL'
0	0	1	0	1
1	1	0	1	0
X	1	1	0	0

b
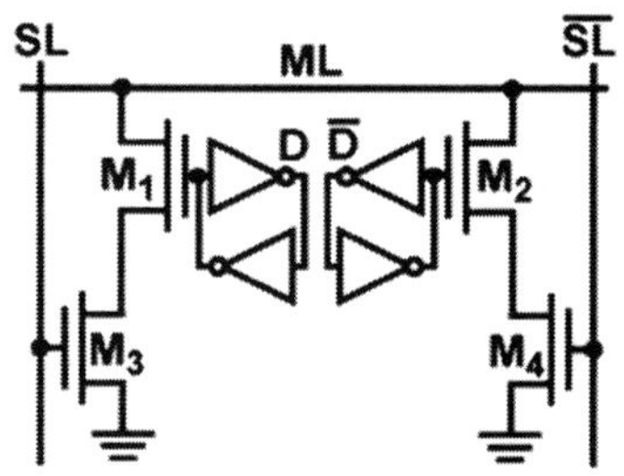

Fig. 14.1 NOR based T-CAM cell implementation and its encoding

14.2.2 Design and Organization of T-CAM

The organization for a given memory bank is shown in Fig. 14.2, which can store up to "n" partitions, where each partition has a maximum of "k" product terms. During each cycle, information stored in the schedule table is used to enable the match lines for one of the "n" partitions stored in a given memory bank. Within each partition, it is possible for more than one match to occur, thereby enabling more than one match line. One of the access transistors for the 6T-SRAM cell is gated by the match line, so that data is read out only if a match occurs. In case a matching PLA entry of a given function has logic value of "1", the node "V" is discharged through the corresponding NOR leg. The drivers for the input data and the current sense amplifiers are shared for each memory bank, thereby reducing the interface logic required. Note that partitions with product terms less than "k" have $D = D_c = 0$, which always leads to a mismatch in the unused match lines.

14.2.3 A Hybrid CAM-LUT Approach

As illustrated in [4], a PLA representation is good for complex inter-related function while LUT based representation as in traditional FPGAs is suitable for small complex gates. Such a scenario is often encountered for circuits that require many "XOR"-like logic function leading to exponentially large number of product terms in the PLA representation. Hence, a framework containing both PLA and LUT based representation is advantageous for memory-efficient realization of all classes of functions. Hybrid CAM/LUT-based storage for the memory based computing framework can therefore be used. The partitioning algorithm was augmented to allow the user to choose a LUT based implementation for the partitions for which a CAM-based implementation of the product terms requires a larger area compared to the corresponding LUT area. Thus, a hybrid approach can potentially improve the area requirement for the total memory to which the application is mapped.

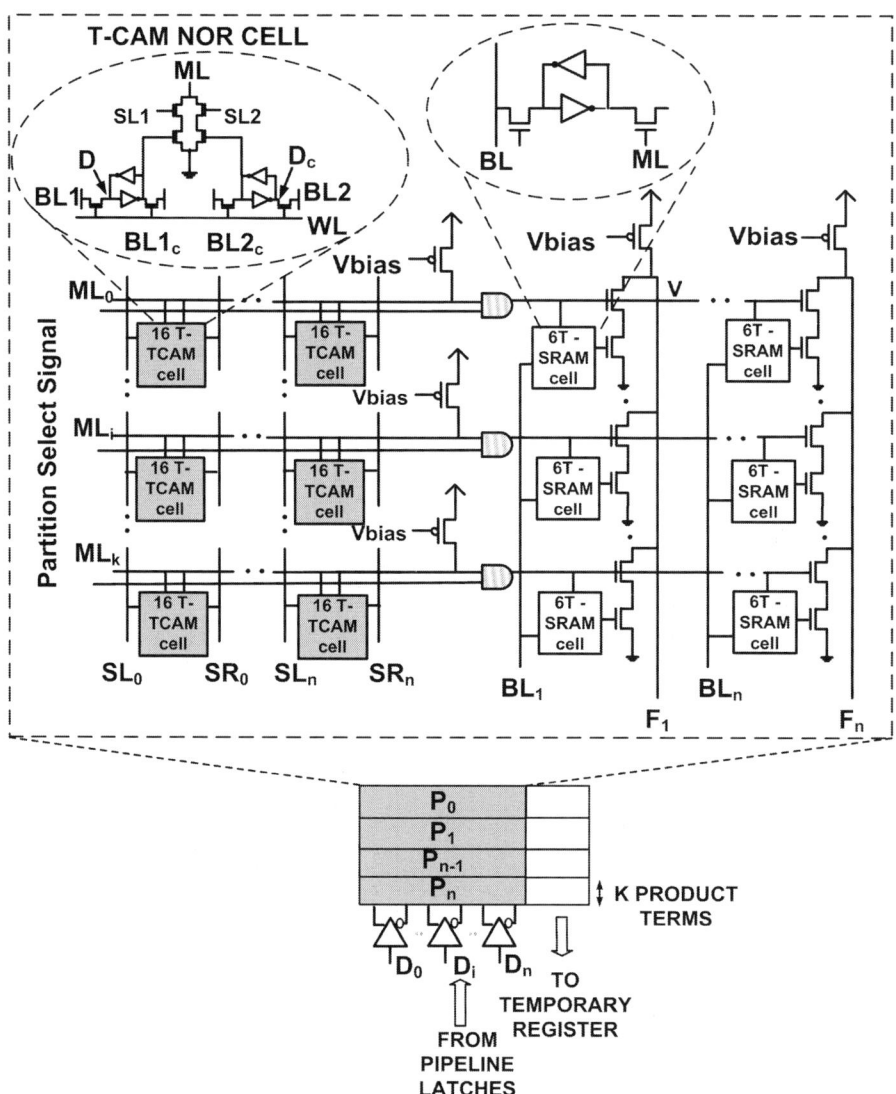

Fig. 14.2 T-CAM based realization of individual partitions mapped to the MBC framework

14.3 Power and Performance of a CAM Based MBC Framework

14.3.1 Delay and Power Estimation

Spice simulations were carried out with PTM 70nm technology model [6] for the CAM-based MLB framework. Simulation results for the example MLB with associative storage indicate that the delay for each stage of the pipeline is within 0.6 ns for four memory banks, each with 16KB of memory. The total power consumption for a cycle time of 0.6ns and input switching activity of 12.5% (default switching activity for Quartus v7.0 simulator by Altera) was estimated to be 10.46 mW.

14.3.2 Improvement in Memory Requirement

Memory requirement was noted for both ISCAS-85 benchmark suite, DCT and FIR applications when mapped to (i) LUT and (ii) CAM based implementation of the multi-cycle computing framework. Table 14.1 compares the memory requirement, delay and energy/vector for the two scenarios. From Table 14.1 and Fig. 14.3a it can be observed that a CAM-based implementation entails an exponential saving in memory requirement compared to a LUT-based approach. It has also been observed that with CAM based implementations, it is possible to map partitions with larger input and output sizes (16×16) with significant savings in area (251X), with nearly the same speed-up as offered by a corresponding LUT based implementation (-7.62%). To observe the impact of hybrid CAM/LUT based application mapping, simulations were performed with several ISCAS-85 benchmarks. Figure 14.3b shows the memory requirement (in KB) and area (estimated as active area, $W \times L$) with the CAM based MBC implementation. As may be observed from Fig. 14.3b, the hybrid approach can further improve the memory saving for some circuits, which consist of partitions with large number of product terms.

Table 14.1 Comparison of memory requirement, area, delay and energy/vector for 12×12 partitions

Ckts	Mem. Req (KB)		Area		Delay (ns)		Energy/Vector(pJ)		Leak Pwr(μW)	
	LUT	CAM	LUT	CAM	LUT	CAM	LUT	CAM	LUT	CAM
C880	37.6	1.11	301	11.4	3.5	3.5	29.57	36.61	2.98	0.49
C1908	20	3.41	160	35.3	2.5	3	21.13	31.38	1.58	1.52
C2670	81.7	1.93	653.6	19.7	5	5	42.25	52.3	6.47	0.85
C3540	130.9	2.68	1047	27.2	8.5	8.5	71.83	88.91	10.36	1.17
C5315	111.4	4.13	891.4	42.3	9	9	76.05	94.14	8.82	1.83
C7552	180.1	7.83	1441.1	80.3	11.5	11.5	97.18	120.29	14.27	3.47
C6288	216.2	12.9	866.4	130.8	9.5	10	80.28	104.6	8.57	5.65
DCT	44	15.83	352	168.85	106	110	895.70	1150.6	3.48	1.67
FIR	46.75	16.76	374	178.83	14.5	16	122.52	167.36	3.71	1.77
Avg Improv.	+22.58X		+16.94X		−4.37%		−29.2%		+61.8%	

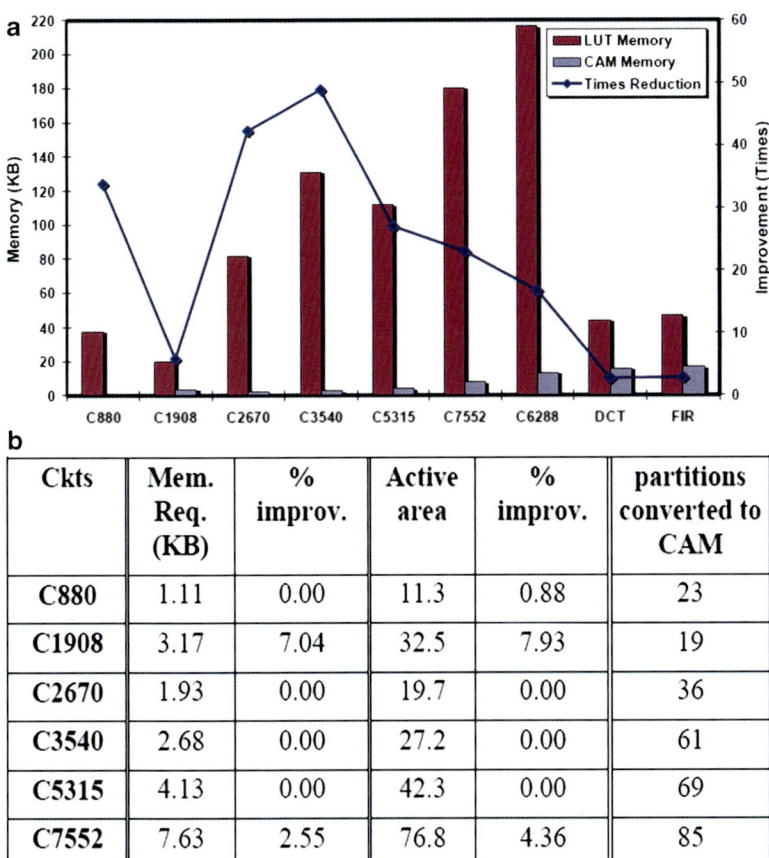

Fig. 14.3 (a) Comparison of memory requirements between LUT and CAM-based implementations; (b) Memory and area requirement with hybrid CAM-LUT implementations

References

1. J. Hutchby, M. Garner, "Assessment of the Potential and Maturity of Selected Emerging Research Memory Technologies". Technical report, ITRS, 2010
2. K. Pagiamtzis, A. Sheikholeslami, "Content-Addressable Memory (CAM) circuits and architectures". IEEE J. Solid State Circuits, 41(3), 712–727 (2006)
3. A. Agarwal et al., "A Dual-Supply 4GHz 13fJ/bit/search 65x128b CAM in 65nm CMOS", in *ESSCIRC*, 2006
4. A. Kaviani, S. Brown, "Hybrid FPGA Architecture", in *Intl. Symp. on FPGAs*, 1996
5. F. Heile, A. Leaver, "Hybrid Product Term and LUT based Architectures Using Embedded Memory Blocks", in *Intl. Symp. on FPGAs*, 1999
6. [Online], "Predictive Technology Model". http://ptm.asu.edu/

Part VI
Off-Chip Hardware Acceleration Using MBC

Chapter 15
Background and Motivation

Abstract This chapter describes the Von-Neumann bottleneck and its effect in limiting the energy-efficiency of traditional compute hardware for processing data-intensive applications. For these applications, traditional approach of bringing data from off-chip memory to on-chip compute engines has been demonstrated to be ineffective in terms of energy-efficiency. In the era of big data, on-chip computing is restricted by both off-chip bandwidth and access energy. In such a scenario, an off-chip compute framework realized by instrumenting non-volatile memory can prove to be an effective solution in mitigating the Von-Neumann bottleneck. This chapter also discusses the nature of such an off-chip compute hardware and the applications which can benefit from it.

15.1 Von-Neumann Bottleneck

The term "Von-Neumann bottleneck" famously coined by John Backus in his 1977 ACM Turing award lecture refers to the limited data transfer rate between the processor and the memory compared to the large amount of memory available [1]. Since the inception of digital computer this bottleneck has been both a performance limiter and more importantly an energy bottleneck for the entire system. Since the transfer rate is not increasing accordingly to Moore's law [2], the relative time required to fetch a volume of data proportionate with the compute resources on-chip is increasing with every technology node [3]. This limitation is exacerbated by the emergence of data-intensive applications not only for scientific computing but from diverse areas of daily life such as web serving, order processing, inventory control, data mining, multimedia storage, searching and characterization [3,4]. For these applications, on-chip hardware devote most of their processing time and energy in moving data rather than in complex calculations. These applications are primarily bound by off-chip input/output (IO) performance and energy requirements. Embedded cache hierarchy partially alleviates this IO bottleneck but have already hit limits in hiding the off-chip access latency [5].

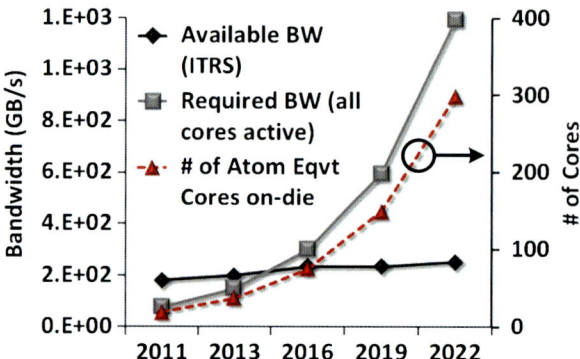

Fig. 15.1 Trends in off-chip BW with technology scaling. The gap between available and required BW is increasing with technology generations. Projected available BW and number of Atom equivalent cores were obtained from [6] and [7], respectively

In case of the data-intensive applications, they contribute heavily to the average energy expended per instruction (50% or more [8]). In such a scenario, doubling of transistor density per chip with the aim of speeding up mathematical operations that are already 100× faster than argument fetch and store is unlikely to bring any benefit in energy-efficiency [3]. This section highlights the performance and power limitation imposed by this back-and-forth data transfer between processor and off-chip memory.

15.1.1 Performance Barrier Due to Von-Neumann Bottleneck

Projections for off-chip bandwidth and on-chip integration density from ITRS [6] for future technology nodes suggest that off-chip bandwidth scales poorly in comparison to on-chip transistor density. While on-chip compute density is likely to improve by a factor of *16X* from 2011 to 2022, the off-chip bandwidth is expected to increase by only *40%* over the same period. Assuming each core running at 1GHz has a 64-bit main memory interface, the worst-case off-chip bandwidth requirement at a given technology node can be estimated as *(# of cores)×(bandwidth per core)*. For each technology node over this time period, the number of Intel Atom equivalent cores that can be accommodated on a single die can be obtained from [7]. From Fig. 15.1a, it may be observed that, if all the cores are simultaneously accessing off-chip memory, then the bandwidth requirement far surpasses the available bandwidth, which can only support *10%* of the on-chip cores. It is worth noting, that off-chip memories have much higher internal bandwidth. For example, if a 2KB page is simultaneously read from each 4,096 blocks of a 45 nm 1GB Flash memory with a read cycle time of 20 ns, then the bandwidth available inside the Flash array is 4.2×10^5 GB/s. In contrast, the bandwidth at the 16-bit Flash interface running at 20 ns, is only 100 MB/s.

15.1 Von-Neumann Bottleneck

$$t_{proc} = \frac{1}{N_{issue}} \times (1+$$

$$g \times (h_{L1} \times L1_{hitlat}$$

$$+ m_{L1} \times (L1_{misslat} + h_{L2} \times L2_{hitlat})$$

$$+ m_{L1} \times m_{L2} \times (L1_{misslat} + L2_{misslat}$$

$$+ h_{LD} \times LD_{hitlat})$$

$$+ m_{L1} \times m_{L2} \times m_{LD} \times (L1_{misslat} + L2_{misslat}$$

$$+ LD_{misslat} + GD_{accesslat}))$$

or

$$t_{proc} = \frac{1}{N_{issue}} \times (1 + g \times T_{mem}) \tag{15.1}$$

$$e_{proc} = E_{compute} + g \times (h_{L1} \times L1_{hiten}$$

$$+ m_{L1} \times (L1_{missen} + h_{L2} \times L2_{hiten})$$

$$+ m_{L1} \times m_{L2} \times (L1_{missen} + L2_{missen}$$

$$+ h_{LD} \times LD_{hiten})$$

$$+ m_{L1} \times m_{L2} \times m_{LD} \times (L1_{missen} + L2_{missen}$$

$$+ LD_{missen} + GD_{accessen}))$$

or

$$e_{proc} = E_{compute} + g \times E_{mem} \tag{15.2}$$

15.1.2 Energy Barrier for Data-Intensive Applications

A detailed study presented in [3] and [9] reveals that *managing latency and energy for the memory and interconnect* is the key to achieving energy-efficiency in future technology nodes. These data-intensive applications either consist of: (i) kernels operating independently on non-overlapping datasets (e.g. encryption/target recognition/filtering etc.) or (ii) kernels working in a collaborative fashion on non-overlapping datasets (e.g. hash generation, mapreduce etc.) to produce the final output. In order to identify the major hurdles to energy scaling, the performance of 10 common kernels has been simulated using the *Simplescalar v3.0d* toolset [10] for the processor configuration provided in Table 15.1. Each of these applications was compiled for the *PISA* instruction [10] set with −O2 option. The compiled binary was then simulated for the input datasets shown in Table 15.3. For each of these

Table 15.1 Processor configuration

Processor	2-way in-order issue, 2 integer ALUs
	2 integer mul/div units, 2 FP ALUs
	2 FP mul/div units, 2 Wr/Rd ports
Branch Prediction	Combined, 32-entry RAS, 2K 4-way BTB, 8 cycle mis-prediction penalty
Caches	32KB 2-way 1 cycle I/D L1, 512KB, 4-way 12 cycle L2
Main memory	100 cycle latency, 8-byte wide bus

Table 15.2 Typical performance and energy values at 45nm node [3]

Performance Parameters	# of Cycles
Compute Latency	1
L1 hit latency	1
L1 miss/L2 hit latency	8
Local DRAM access latency	100
Global DRAM/non-volatile memory access latency	10,000
Energy Parameters	**Value (pJ)**
Compute	20
DL1/IL1 Hit	40
L1 miss/L2 hit	400
Local DRAM access	1,400
Global DRAM/non-volatile memory	14,000

kernels, key system-level performance metrics such as cache hit/miss frequency etc. were noted. Values for these metrics were combined with representative latency and energy values at 45nm technology node (refer to Table 15.2) to estimate the system-level energy and latency requirements (Table 15.3). In presence of on-chip caches, effective number of cycles per instruction (t_{proc}) for a processor can be computed using (15.1). In (15.1), h_{L1}, h_{L2}, h_{LD} and h_{GD} represent the hit percentage in L1, L2, local DRAM and global DRAM respectively. m_{L1}, m_{L2}, m_{LD} and m_{GD} are respective miss percentages. $L1_{hitlat}$, $L1_{misslat}$, $L2_{hitlat}$, $L2_{misslat}$, $localDRAM_{hitlat}$, $localDRAM_{misslat}$ and $globalDRAM_{accesslat}$ stand for the hit/miss latencies associated with L1, L2, localDRAM and access latency for globalDRAM respectively. In (15.1), "g" represents the fraction of instructions with memory references and N_{issue} represents the number of instructions issued in a given clock cycle. Similarly, the average energy expended per instruction (e_{proc}) in a software execution model can be expressed as (15.2). In (15.2), $E_{compute}$ represents the energy expended in the processor due to instruction fetch, register file read and in the execution units. $L1_{hit/missen}$ and $L2_{hit/missen}$ represent the L1 and L2 cache hit/miss energies, $LD_{hit/missen}$ and $GD_{accessen}$ represent the hit/miss and access energies for local and global DRAM respectively.

It is important to note that the above derivation is based on the following observations:

15.1 Von-Neumann Bottleneck

Table 15.3 Energy breakdown for a conventional processor executing common computational kernels

Benchmark	Input Size (Byte)	Inst. Cnt	IL1 Access Cnt	DL1 Access Cnt	Total Energy (pJ)	Delay (s)
Advanced Encryption Standard (AES)	2.62E+05	2.33E+09	2.36E+09	1.55E+09	1.99E+11	3.78
Automatic Target Recognition (ATR)	2.62E+05	1.30E+10	1.31E+10	5.44E+09	9.82E+11	21.37
Secure Hash Algorithm (SHA)-1	6.24E+05	2.29E+08	2.30E+08	1.02E+08	1.76E+10	0.36
Motion Estimation (ME)	2.62E+05	6.5E+09	6.64E+09	3.31E+09	5.19E+11	11.97
Smith Waterman (SW) sequence alignment	5.24E+05	8.90E+07	8.91E+07	3.98E+07	7.51E+09	0.16
2D-Discrete Cosine Transform(DCT)	1.05E+06	4.11E+10	4.23E+10	1.18E+10	2.93E+12	53.12
Discrete Wavelet Transform	1.05E+06	4.79E+08	4.82E+08	1.94E+08	3.74E+10	0.68
Max. Entropy Classification (MEC)	2.10E+06	8.51E+08	8.53E+08	3.62E+08	6.46E+10	1.17
Color Interpolation	1.05E+06	1.56E+09	1.57E+09	6.58E+08	1.20E+11	2.33
Census (MapReduce example)	2.10E+06	1.23E+09	1.25E+09	4.42E+08	9.15E+10	1.69
Average		26%	51%	22%		

(i) Data-intensive applications which process peta or terabytes of data typically involve partitioning the data into smaller volumes, each of which can be processed using a single or cluster of processors. In these simulations a lightweight processor processes data of ~ 1 MB size. These simulations capture the off-chip bottleneck for a single processor and evaluates how offloading the task to an off-chip accelerator can potentially benefit one or more processors at each cluster node.

(ii) Experiments have also been carried out with larger data sets (~ 100 MB). On-chip cache miss rates and % contribution of each component to the system energy was however found to reach an asymptotic value for the data set sizes indicated in Table 15.3.

From Table 15.3, it may be noted that on an average, *73%* of the total energy expended in the processor is contributed by access to on-chip instruction and data caches. Only *26%* of the total energy is actually invested in useful computations, including fetch and decode operations. Although the absolute cache contribution to total power consumption will vary based on the selected processor configuration, buffering of off-chip data in on-chip caches will lead to substantial energy wastage for data-intensive tasks. Among the kernels investigated, *Census* exhibits maximum off-chip access.

15.2 Mitigating Von Neumann Bottleneck Through In-Memory Computing

There is no silver-bullet solution which caters to all data-intensive applications. However, the fact that almost 75% of the energy in a typical processor is dissipated in data transport (from off-chip memory to on-chip memory and between the compute pipeline and on-chip memory), suggests that optimizing the compute model for data-intensive tasks can reap large improvements in energy-efficiency. Note that this bears two implications for the compute model:

(i) physically relocate compute resources closer to the last level of non-volatile storage which drastically minimizes the overhead for data transfer to on-chip execution units.

(ii) replace the conventional software pipeline and caches with a distributed compute and memory infrastructure which leverages local computation to minimize memory and interconnect power dissipation.

In this context, a memory-centric hardware accelerator (MAHA) which interfaces with the last-level non-volatile storage and allows direct offloading of data-intensive tasks can dramatically improve energy-efficiency through a close association of memory and compute logic.

15.2.1 Applications Amenable to In-Memory Computing

Before mapping a kernel to an in-memory accelerator, key application and system primitives can be leveraged to determine whether it will benefit from in-memory acceleration.

15.2.1.1 Key Primitives

In order to compare between a software-only solution and a hybrid system with off-chip in-memory accelerator, the application characteristics and the system configuration are expressed using a set of primitives as listed below:

1. g—fraction of total instructions with memory reference (loads and stores).
2. f—fraction of total instructions transferred to an off-chip compute engine.
3. c—fraction of instructions translated from host's ISA to the ISA for the off-chip compute framework. Note that loads and stores can be partially removed during such a translation.
4. o—fraction of original instructions which result in an output. A fraction $f \times c \times o$ thus produces outputs which needs to be transferred to the host processor.
5. $e_{offchip}$—average energy per instruction in the off-chip compute engine.
6. e_{txfer}—energy expended in the transfer of an output from the off-chip framework to the host processor.
7. $t_{offchip}$—ratio of cycle time for the off-chip compute framework to that for the host processor.
8. n—fraction speedup due to parallelism in the framework (ratio of on-chip to off-chip compute engine count).
9. t_{txfer}—time taken in terms of processor clock cycles to transfer an output from the off-chip compute framework to the host processor.

With these primitives, the average time to execute an instruction in a system with a host processor and the off-chip compute framework can be formulated as:

$$T_{sys} = T_{offchip} + T_{proc} + T_{txfer} \tag{15.3}$$

where

$$T_{offchip} = t_{offchip} \times (I_{f>g}(f)) \times (f - g + f \times c \times n)$$
$$+ I_{f \leq g}(f) \times f \times c \times n)$$
$$T_{proc} = (1 - f) \times t_{proc}$$

Table 15.4 Typical values of performance and energy in an off-chip compute framework

$e_{offchip}$	50pJ (higher energy per instr. considering local memory load and store)
e_{txfer}	10,000pJ (same order of magnitude as the energy for an off-chip non-volatile memory access)
$t_{offchip}$	15 (typical processor (@ 1GHz) to off-chip memory cycle times)
n	0.01 (with the assumption that large # of compute engines can be accommodated into large last level memory)
t_{txfer}	10,000 (same order of magnitude as latency for an off-chip non-volatile memory access)

$$T_{txfer} = t_{txfer} \times (I_{f>g}(f) \times ((f-g) \times o + f \times c \times o)$$
$$+ I_{f \leq g}(f) \times f \times c \times o)$$

where

$$I_A(x) = \begin{cases} x \text{ if } x \in A \\ 0 \text{ if } x \notin A \end{cases}$$

In (15.3), $T_{offchip}, T_{proc}$ and T_{txfer} denote the fraction latencies in the off-chip compute framework, due to processor execution and in the transfer of the resultant output from the off-chip platform to the processor. As seen from (15.3), the first and the third terms increase with f, indicating the increase in both $T_{offchip}$ and T_{txfer}. The increase is more prominent once the fraction of task f transferred is more than the fraction g dominated by memory references. $T_{offchip}$ reduces as the number of parallel off-chip computing units increases (n reduces). T_{proc} decreases with increasing f since it has to share less burden of computation. A similar expression for the energy of the resultant system is given below which shows the transfer energy increasing and the processor energy decreasing with increasing f:

$$E_{sys} = E_{offchip} + E_{proc} + E_{txfer} \quad (15.4)$$

where

$$E_{offchip} = e_{offchip} \times (I_{f>g}(f) \times (f - g + f \times c) + I_{f \leq g}(f) \times f \times c)$$
$$E_{proc} = (1-f) \times e_{proc}$$
$$E_{txfer} = e_{txfer} \times (I_{f>g}(f) \times ((f-g) \times o + f \times c \times o)$$
$$+ I_{f \leq g}(f) \times f \times c \times o)$$

With performance and energy values typical to the operation of an off-chip compute framework (modeled after the *MAHA* platform as listed in Table 15.4), the system-level improvement in energy-efficiency is estimated for applications

15.2 Mitigating Von Neumann Bottleneck Through In-Memory Computing

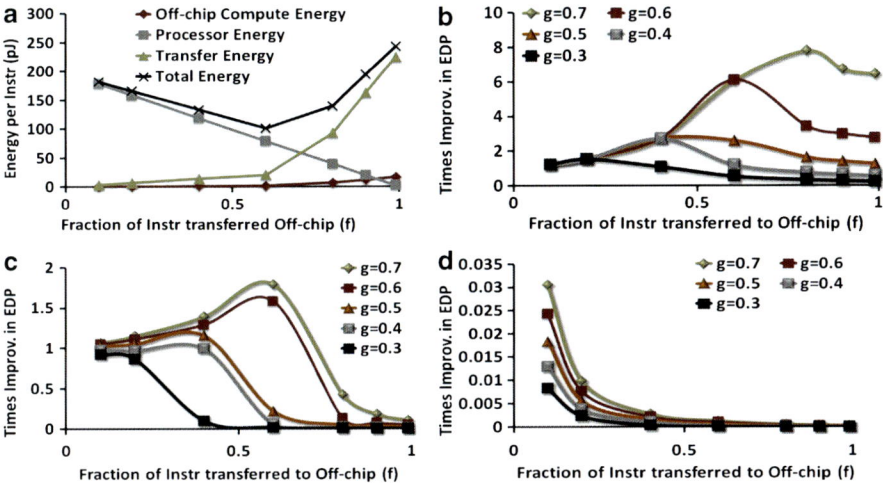

Fig. 15.2 Energy and performance for a hybrid system with a host processor and off-chip memory based hardware accelerator. (**a**) Effective energy per operation in the hybrid system with $c=o=0.05$ and $g=0.7$. Improvement in energy-efficiency (EDP) for the hybrid system with (**b**) $c=o=0.005$; (**c**) $c=o=0.05$; (**d**) $c=o=0.5$

with varying values of the primitives listed earlier. Figure 15.2a shows the three components and the total system energy with $g=0.7$ and $c=o=0.05$, respectively. As clearly evident from Fig. 15.2a, for the values of c and o selected, minimum energy consumption is achieved for values of f close to g. For values of c and o order of magnitude large or small, the total energy was found to always increase or decrease with f, respectively, suggesting that small and large c and o value always favor or disfavor off-chip acceleration. A similar dependance on c, o and f was observed for total execution latency. Combining the performance and energy trends for this system, the EDP trend of the total system (Fig. 15.2b–d) is derived. As expected, the EDP improvement progressively diminishes as the data-intensive nature (value of g) of the application is reduced. Finally, from Fig. 15.2b and c it may be noted that for EDP improvement > 1, for a given g, there exists an f for which maximum EDP improvement can be observed. Given the nature of an application (g and o known), knowledge of its mapping to the off-chip framework (c and n known), it is possible to determine f for which maximum energy savings can be obtained by off-chip computing. From the analysis above, it may be inferred that applications with large number of memory references (high value of "g") and small output data set (low value of "o") are particularly amenable to improvement in energy-efficiency through off-chip computing.

References

1. J. Backus, "Can programming style be liberated from the von-neumann style? A functional style and its algebra of programs". ACM Comm. **21**(8), 613–641 (1978)
2. E.S. Chung, P.A. Milder, J.C. Hoe, K. Mai, "Single-Chip Heterogeneous Computing: Does the Future Include Custom Logic, FPGAs, and GPGPUs? in *Intl. Symp. on Microarchitecture*, 2010
3. P. Kogge et al., "ExaScale Computing Study: Technology Challenges in Achieving Exascale Systems". http://www.cse.nd.edu/Reports/2008/TR-2008-13.pdf
4. V.W. Lee et al., "Debunking the 100X GPU vs. CPU Myth: An Evaluation of Throughput Computing on CPU and GPU", in *Intl. Symp. on Computer Architecture*, 2010
5. R. Murphy, A. Rodrigues, P. Kogge, K. Underwood, "The Implications of Working Set Analysis on Supercomputing Memory Hierarchy Design", in *Intl. Conf. on Supercomputing*, 2005.
6. "International Technology Roadmap for Semiconductors" http://www.itrs.net/links/2009ITRS/Home2009.htm.
7. E.S. Chung, P.A. Milder, J.C. Hoe, K. Mai, "Single-Chip Heterogeneous Computing: Does the Future Include Custom Logic, FPGAs, and GPGPUs?", in *Intl. Symp. on Microarchitecture*, 2010
8. A. Sodani, "Race to Exascale: Opportunities and Challenges". http://www.microarch.org/micro44/files/program.htm
9. P.M. Kogge. "From Petaflops to Exaflops", in *Intl. Supercomputing Conference*, 2008
10. "Simplescalar Toolset v3.0" http://www.simplescalar.com/

Chapter 16
Off-Chip MAHA Using NAND Flash Technology

Abstract In this chapter, details of the hardware architecture for an off-chip *MAHA* framework based on CMOS-compatible Single Level Cell (SLC) NAND Flash memory array (Park et al. A 45nm 4Gb 3-Dimensional Double-Stacked Multi-Level NAND Flash Memory with Shared Bitline Structure, Intl. Solid-State Circuits Conference, 2008) are presented. CMOS-compatibility allows the integration of MLB controller (including registers, datapath and PI) realized using CMOS logic with the Flash process. Flash memory has seen an astounding increase in integration density over the last few years (Technologies for Data-Intensive Computing, http://www.hpts.ws/session1/bechtolsheim.pdf), making it attractive storage system for data-intensive computing (Technologies for Data-Intensive Computing, http://www.hpts.ws/session1/bechtolsheim.pdf; Kgil et al. Improving NAND Flash Based Disk Caches, Intl. Symp. on Computer Architecture, 2008). Although Multi-Level Cell (MLC) Flash has gained more popularity due to its high-integration density and low cost per bit, the consideration for SLC Flash is mainly driven by the availability of opensource area, power and delay models for the same (Mohan et al. FlashPower: A detailed power model for NAND flash memory, DATE, 2010). The proposed architecture, however, applies to MLC Flash memory as well.

16.1 Overview of Current Flash Organization

The study on different off-chip accelerator configurations presented in the previous chapter suggests that:

1. For maximum energy-efficiency, the off-chip accelerator must seamlessly integrate with the off-chip NVM with minimal impact on memory density and performance.
2. The accelerator framework must be capable of exploiting the high internal data bandwidth of the NVM. This requires a parallel compute fabric with low design overhead favorable for mapping arbitrary data-intensive kernels.

Table 16.1 Flash organization and performance

Technology	45 nm
Total size	1 GB SLC
Organization	2,048 bytes * 128 pages * 4,096 blocks
Cell size	$5 * F^2 = 0.010125\,\mu m^2$
Read cycle time	20 ns
Program time	800 µs

3. A spatio-temporal computing model must be leveraged to minimize the PI overhead for the off-chip framework.

To demonstrate the viability of such an off-chip in-memory accelerator, NAND Flash memory, which is the current standard for high-performance secondary memory has been considered in this work. The choice of NAND flash technology is guided by its CMOS-compatibility (which allows integration of static CMOS logic alongside non-volatile memory (NVM)); non-volatility (that justifies its usage as LLM); high integration density; and acceptable read access energy/performance. This off-chip framework, however, can be extended to other emerging non-volatile memory technologies, such as phase change memory (PCM) and spin torque transfer random access memory (STTRAM). A typical organization of the NAND Flash memory is shown in Fig. 16.1a [1] with Flash memory array and a number of logic structures responsible for controlling the read and write operations to the Flash [2]. The Flash Translation Layer (FTL) converts the logical address of a location to its corresponding physical address. The command register latches the operation request commands and status registers are used to indicate the status of operations inside Flash memory. The write buffer is used to store data before they are written into the Flash memory. The NAND Flash typically has 8-bit or 16-bit I/O bandwidth. NAND Flash is organized in units of pages and blocks. Typical page size is 2KB [1] and each block can have 64–128 pages. During normal Flash read, the block decoder first selects one of the blocks, which is followed by the page decoder selecting one of the pages of the block. Contents of the entire page is first read into the page register and then serially transferred to the Flash external interface. Specifications for the baseline Flash organization are provided in Table 16.1.

Fig. 16.1 (a) Modifications to conventional Flash memory interface to realize *MAHA* framework. A small control engine (CE) outside the memory array is added to initiate and synchronize parallel operations inside the memory array; (b) Modified Flash memory array for on-demand reconfigurable computing. The memory blocks (called MLB) are augmented with local control and compute logic to act as a hardware reconfigurable unit. (c) Hierarchical interconnect architecture to connect a group of MLBs

16.1 Overview of Current Flash Organization

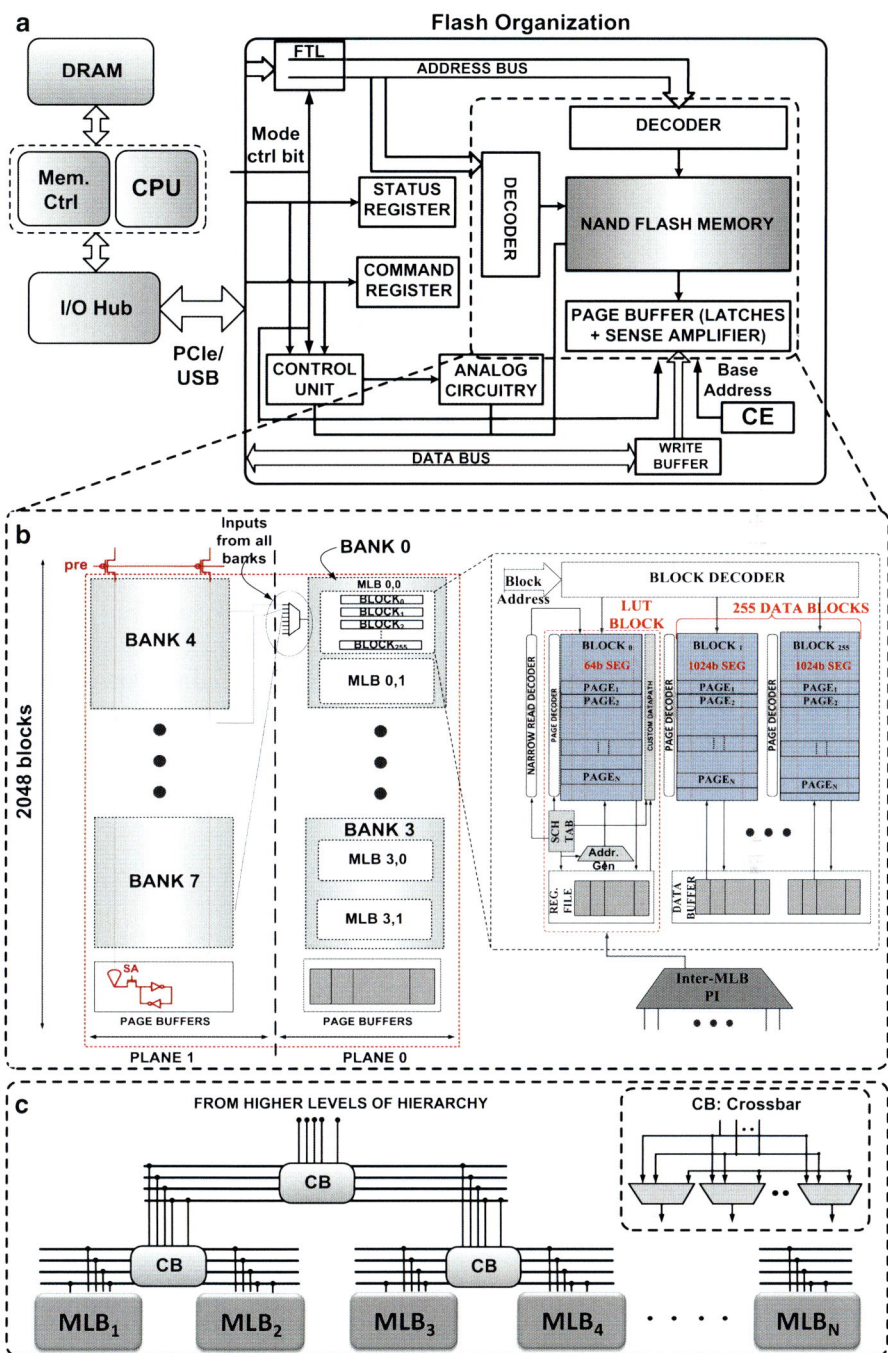

16.2 Modifications to Flash Array Organization

In this section, design modifications to instrument the Flash memory array in order to achieve on demand computation, without affecting normal memory read/write operations are considered. These design changes allow the Flash memory to essentially transforms into an array of MLBs, which communicate over a hierarchical interconnect.

16.2.1 Compute Logic Modifications

A group of N Flash blocks is logically clustered to form a single MLB. In each MLB, the blocks are logically divided into LUT blocks (storing LUT responses) and data blocks (storing operands). Ratio of data to LUT blocks impacts both the area overhead and the flexibility to map a large application. Inside each MLB, operations are either mapped as LUT or custom datapath operations. MLB control logic (i.e. schedule table) and custom datapath are implemented using static CMOS logic. Outputs from each operation that are not consumed in the next clock cycle have to be stored and retrieved later. In light of write endurance problems, use of the Flash array for temporary data storage is avoided and a custom *dual ported asynchronous read register file* is used for storing the intermediate outputs. Input operands for the LUT or the datapath may come from either the register file or another MLB. A fast intra-MLB multiplexor tree consisting of pass gate multiplexors and weak keepers is responsible for selecting appropriate operand(s) for the LUT and the datapath. All operations within a given MLB are scheduled beforehand and stored as a μ-code inside the schedule table, implemented using a 2-D flip-flop array.

For a normal NAND Flash read, the entire page (typically 2KB) is read at a time. However, for LUT operations, due to smaller operand sizes (16 bits), a wide read is avoided. To achieve this, a *narrow-read* scheme is employed (Fig. 16.2) for the LUT blocks in which a fraction of a page size is read at a time. This scheme, however, incurs hardware overhead due to wordline segmentation. To minimize this overhead, only 64-bit words are read from each block at a time. The main advantage of this scheme is that it improves energy efficiency by lowering the wordline capacitance and bitline to be driven—only 64 Flash cells have to be driven by a wordline and precharging is minimized to 64 bitlines (in the worst case). The combinational logic required to switch between a narrow-read for *MAHA* operation and full page read for normal Flash operation can be moved outside the cell array and used together with the narrow read decoder to control the AND gates used for segmentation. In order to exploit the data locality inside a page, the segmentation for data blocks is coarse with data sizes of 4096 bits being read out from each page and stored inside buffers. A group of such LUT and data blocks constitutes one MLB, as shown in

16.2 Modifications to Flash Array Organization

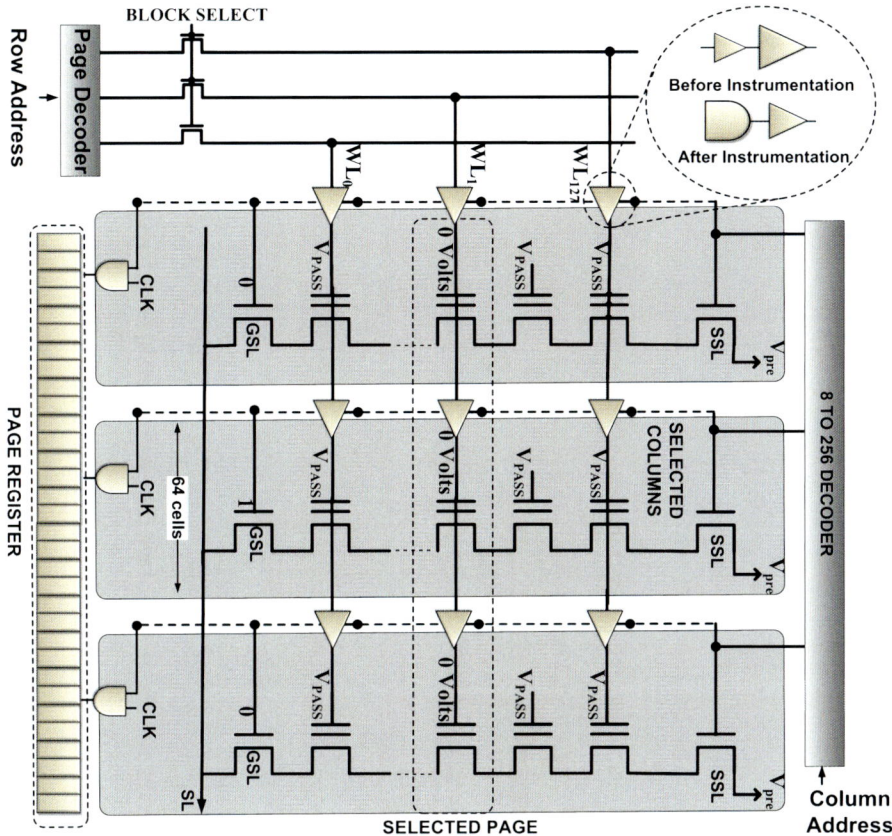

Fig. 16.2 Modifications to the Flash array architecture to support narrow read operation. The 8to256 decoder is responsible for selection of a set of columns from the entire page size

Fig. 16.1b. The two planes of the Flash array are logically divided into 8 banks each consisting of 2 MLBs each. Each MLB contains 256 blocks of Flash memory and comprises of 1 LUT block and 255 data blocks.

16.2.2 Routing Logic Modifications

In the baseline Flash organization, each block communicates with the page register over a shared bus. In order to minimize the inter-MLB PI overhead, a set of hierarchical buses is assumed in this architecture with a crossbar present at each level of hierarchy to select the source of the incoming data (Fig. 16.1c). The hierarchy consists of 4 levels similar to banks, subbanks, mats and subarrays present in a typical cache data array [3].

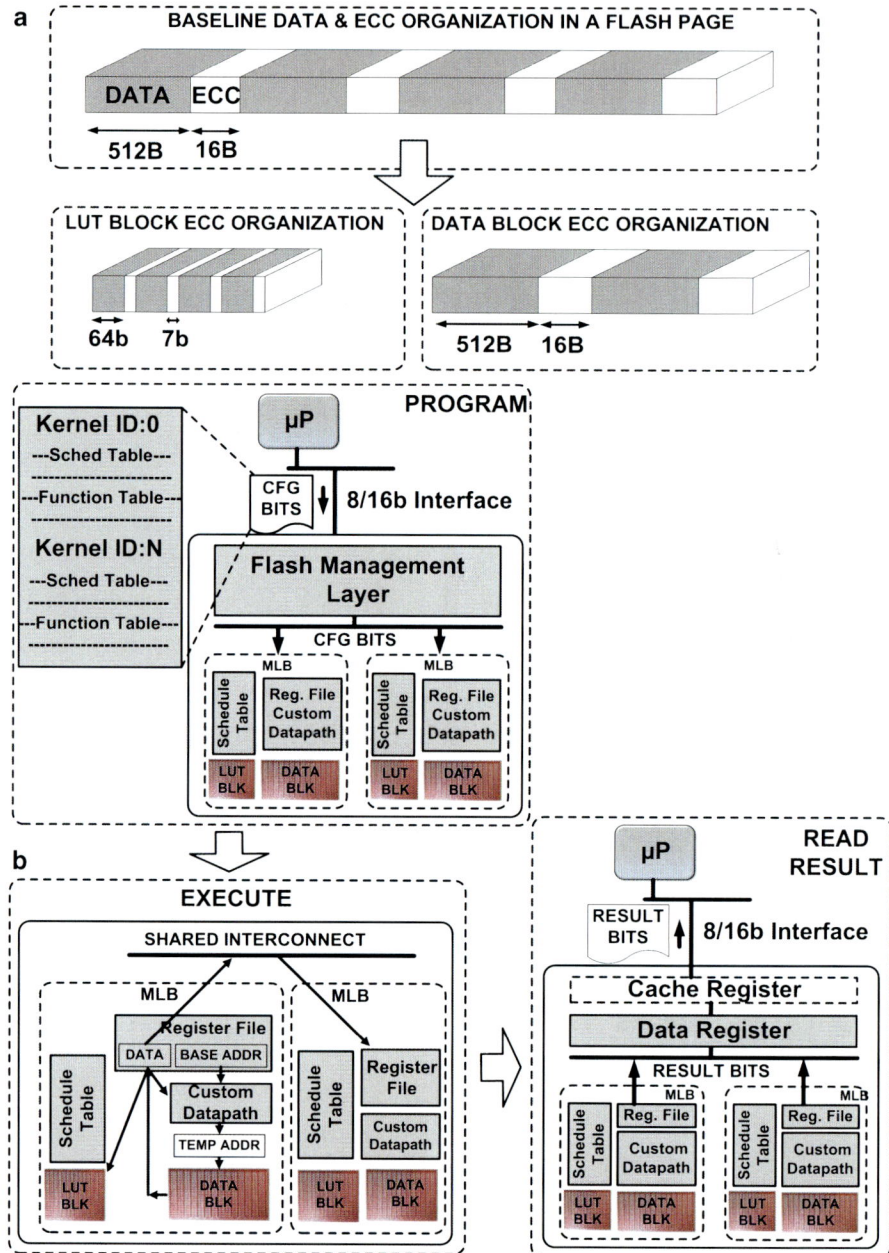

Fig. 16.3 (a) ECC organization in conventional and modified Flash memory; (b) Three different phases of *MAHA* operation

16.3 ECC Computation and Modes of Operation

Block management and Error Control Coding (ECC) are well-known techniques to ensure reliability for normal Flash read and write operations. Typically ECC in NAND Flash is capable of correcting 1-bit and detecting up to 2-bit errors (SECDED) per 512 bytes [4] and ECC bits are interleaved with data as illustrated in Fig. 16.3a. With selection granularity of 512B in data blocks, no additional ECC bits are required for these blocks. ECC overhead is incurred only for LUT blocks with a selection granularity of 64-bits. For the SECDED scheme, 7 parity bits are interleaved with each 64 bit LUT response as illustrated in Fig. 16.3. This leads to almost 3X increase in ECC storage per LUT block. However, large ratio (255:1) of data to LUT blocks minimizes the overall ECC storage overhead to less than 1%. ECC check is performed for each data and LUT read operation. Any error detected by SECDED hardware is indicated to the Flash management layer and execution is stalled.

MAHA framework has 3 phases of operation:

i. *PROGRAM:* The application mapping tool runs on the host processor generating a binary image file which is transferred to the Flash management layer. The Flash management layer is conventional FTL with additional service modes to support *MAHA*. The management layer is responsible for distributing the configuration bits to individual MLBs and asserting the execution start signal. The configuration consists of μ-code for schedule tables and LUT responses.

ii. *EXECUTE:* Each MLB begins executing the μ-code in its schedule table when execute signal is asserted. Data is accessed from the data blocks using an address obtained by adding an offset to the base address. The offset is incremented in each iteration. Operands from data block serve as an input to either a LUT block or to the custom datapath. All operations are scheduled and execute in a topological fashion.

iii. *READ RESULT:* Total duration of EXECUTE phase is known through scheduling and READ RESULT phase begins on expiration of that duration. Resultant data from individual MLBs are captured in the data register in the EXECUTE step. This resultant data is then transferred from the data register to the cache register and subsequently to the host processor. The READ RESULT phase is overlapped with another EXECUTE phase.

References

1. K.T. Park et al., "A 45nm 4Gb 3-Dimensional Double-Stacked Multi-Level NAND Flash Memory with Shared Bitline Structure", in *Intl. Solid-State Circuits Conference*, 2008
2. [Online]. "Micron 1Gb NAND Flash Data Sheet". http://download.micron.com/pdf/datasheets/flash/nand/1gb_nand_m48a.pdf
3. "Cacti 6.5". http://www.cs.utah.edu/~rajeev/cacti6/cacti6-tr.pdf

4. "Bad Block Management and ECC in Micron® Single-Level Cell (SLC) NAND" http://www.micron.com/products/support/technical-notes
5. V. Mohan, S. Gurumurthi, M.R. Stan, "FlashPower: A Detailed Power Model for NAND Flash Memory", in *DATE*, 2010
6. "Technologies for Data-Intensive Computing" http://www.hpts.ws/session1/bechtolsheim.pdf

Chapter 17
Improvement in Energy-Efficiency with Off-Chip MAHA

Abstract This chapter first arrives at an optimal architecture for an off-chip MAHA framework based on SLC NAND Flash technology. The viability of mapping diverse applications to this framework is studied and the energy-efficiency of the mapped applications is compared against a baseline model, i.e. a software based execution with no acceleration. Next the efficiency of this off-chip accelerator with commercial FPGA and GPU based acceleration. Finally a details of an emulation setup which validates the functionality of the MAHA framework is presented. Both simulation and emulation results presented in this chapter confirm that an off-chip in-memory accelerator can reap considerable gains in energy-efficiency for data-intensive tasks.

17.1 Design Space Exploration for Off-Chip *MAHA*

Table 17.1 lists the key design parameters for the off-chip *MAHA* framework. For the applications selected, the MBC software flow explores each of these parameters independently and arrives at the final *MAHA* configuration based on performance and resource constraints provided. Steps followed for design space exploration are:

1. Based on the area/delay/energy values for individual MLB components and the routing switches, estimate design overhead for entire MLB as well as for inter-MLB PI at each point in the design space. The total number of design points considered for the Flash based MAHA framework is 27,648.
2. Map the benchmark applications to the MAHA framework and obtain a set of viable configurations for which all the benchmarks can be mapped.
3. Calculate the area overhead, performance and energy requirements for the set of viable configurations and select the best configuration (design point) based on the primary design parameter (i.e. energy-efficiency).

Table 17.1 *MAHA* design parameters

MLB Parameters	Comments
Granularity	Minm # of bits that can be uniquely selected inside MLB
Datapath Width	Width of adder, multiplier is half the adder size
Max MLB Level	Maxm # of consecutive vertex levels which can be accommodated inside a MLB
MLB In/Out	# of inputs/outputs to each MLB
Scheduling Parameters	**Comments**
Max Parallel Datapath op	Max # of parallel datapath operations allowed inside MAHA framework
Max Parallel LUT op	Max # of parallel LUT operations allowed inside MAHA framework
MAHA hierarchy Parameters	**Comments**
Banks, Subbanks Mats and Subarrays	# of elements at each level of a hierarchical memory organization
Buswidth	Interconnect bandwidth at input of each bank # of inputs to subbanks, mats and subarrays are progressively reduced based on the # of elements in each level of hierarchy

System level energy and performance results are then calculated based on the total size of data being processed and energy and latency values for a single operation. The reconfiguration overhead for each application is also estimated.

17.1.1 Energy, Performance and Overhead Estimation

The cycle time of 20 ns for MAHA operation is calculated from the critical path delay components, namely bitline precharge time (12 ns), intra-MLB delay (3 ns) and inter-MLB signal propagation time (5 ns). Logic components were synthesized using 45 nm standard cell library for minimum energy constraint. Area, delay and energy models for CMOS compatible SLC NAND Flash memory were obtained from existing literature [1, 2]. These were used to estimate the design overhead for the narrow read scheme presented in the previous chapter.

1. *Overhead for custom datapath*: The logic datapath was synthesized for width—8, 16 and 32 bits.
2. *Overhead for schedule table:* Two design parameters relevant to the schedule table are: (i) value of "N" in the modulo-N scheduling policy; (ii) number of parallel operations per cycle inside each MLB.
3. *Overhead in Flash array for narrow-read operation:* The area for a single block of the Flash array is modeled using the equation $5 * F^2 * (Npages) * (pagesize)$.

17.1 Design Space Exploration for Off-Chip *MAHA* 167

Table 17.2 Flash array energy break-up for normal and MAHA modes

Parameter	Normal operation	MAHA LUT	MAHA Data
Ebl,pre (pJ)	54309.86	0.02	338.1
Ewl,pre (pJ)	424	85	106.07
Esel (pJ)	18.73	0.67	1.6
Eunsel (pJ)	2261.9	81.8	193.1
Esl,r (pJ)	0.025	0.00125	0.00228
Ebl,r (pJ)	27154	0.0135	169
Esense (pJ)	44.23	0.1728	11.76
Edec (pJ)	377	135.6	141.24

For the baseline Flash array specifications provided in Chap. 16, this area at 45 nm technology node is calculated to be 21,234 μm². In order to support narrow read operation, an additional 8to256 decoder occupying 763 μm² is inserted to select 64 bit output from the page selected by the page decoder. Since the LUT block is separate from the the data blocks (which reads 4,096 bits at a time), area overhead for the two cases are different. For the LUT block the area overhead is 1,327 μm², 680 μm² and 18 μm² respectively, while for the data blocks the area overhead is 1,327 μm², 21μm² and 1,161 μm², respectively.

Energy consumed in the Flash array is modeled following the FLASHPOWER model as presented in [1]. Energy components for a read operation include energy of the decoders ($Edec$), bitline (Ebl, pre) and wordline precharge (Ewl, pre) energy, energy for driving the selected ($Esel$) and unselected pages ($Eunsel$), energy for turning on the select transistors (Esl) and sense amplifier energies ($Esense$). Table 17.2 provides a comparison of the normal Flash read energy and *MAHA* Flash read energy. The difference in the energy values between the two modes is due to the narrow read scheme and bitline segmentation for the MLB blocks. It is important to note that even though data block read (narrow read=4,096) consumes much more energy than LUT block read, the former is infrequent than the latter.

17.1.2 Selection of Optimal **MAHA** Configuration

As design parameters for the *MAHA* framework were varied, following metrics were noted: (i) area overhead (refer to Fig. 17.1a); (ii) latency (in terms of clock cycles) for a single iteration; (iii) # of MLBs required to map the application; (iv) total energy dissipated in the MLBs; (v) area and energy for inter-MLB PI; and (vi) size of reconfiguration data. *Geometric mean* of these metrics is first calculated across all benchmarks and the final configuration (refer to Table 17.3) is selected on the basis of: (i) *area overhead (determines datapath size and "N" in modulo-N scheduling)*;

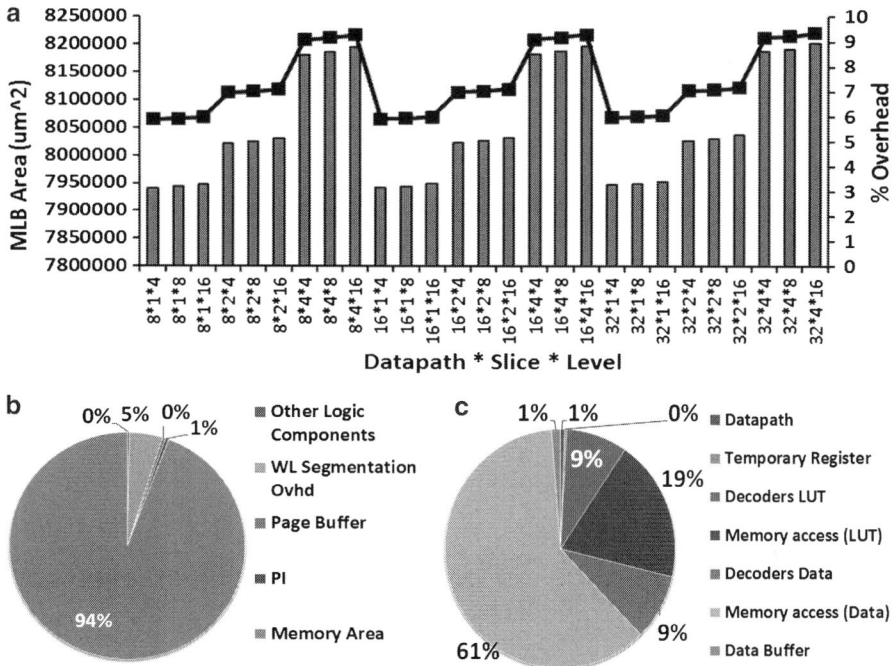

Fig. 17.1 (a) Area for a modified memory block for different values of the design parameters and the corresponding overhead w.r.t an non-modified one; (b) Relative contribution of different components to the total area of the modified Flash; (c) Relative contribution of memory and logic components to the total energy consumed in each MLB during MAHA operation

Table 17.3 Final *MAHA* hardware configuration

Final Configuration	
Hierarchy Parameters	Banks: 8, Subbanks: 1, Mats: 1
	Subarrays: 2, Buswidth: 256
MLB Parameters	Datapath: 16, Op. per cycle:1 Levels: 16
MLB Area	7,913,925 μm^2
Interconnect Area	35,891 μm^2

(ii) *performance (determines # of parallel op inside each MLB.)* Table 17.3 reports the area for a modified memory block and the total interconnect area corresponding to *buswidth=256* for a group of 16 MLBs arranged in a hierarchy of *8,1,1,2*. Since each MLB comprises of 256 blocks, all 4,096 blocks of the 1 GB Flash are organized into the 8,1,1,2 hierarchy. Area overhead for this final *MAHA* configuration stands at only 6%. Figure 17.1b and c provide the area and energy breakdown for this final *MAHA* configuration.

17.2 Energy and Performance for Mapped Applications

Table 17.4 shows the mapping results for a single CDFG instantiation for each of the selected benchmarks mapped to the final *MAHA* hardware configuration. The latency as reported by the tool is obtained from the vertex levels in the critical path of the CDFG and the time period @ 20 ns. During mapping, one or more sections of the input CDFG was folded to maximize hardware reuse, resulting in an effective latency of $Latency \times Degree\ of\ folding$. Based on energy and performance estimates for a single iteration, the same is estimated for a larger dataset, taking into consideration the reconfiguration overhead. Finally, this result is combined with the latency and energy values required to transfer the resultant output from off-chip memory to the host processor to arrive at system-level energy and performance results. From Table 17.4, it may be noted that for *MAHA* the average PI energy is significantly ($10.3X$) less compared to the average MLB logic energy.

17.2.1 Comparison with a Conventional GPP

Energy and performance of a hybrid system with both software execution engine and *MAHA* hardware accelerator is compared against a software only solution with the assumption that the resultant output from *MAHA* framework is transferred back to the main memory. Such a hybrid framework was realized using the Simplescalar v3.0d toolsuite [3] with the baseline processor configuration outlined in Table 15.1. Table 17.5 shows the average contribution of each component to the total energy/delay for the hybrid system.

Table 17.4 Mapping results for a single input vector with final *MAHA* hardware configuration

Benchmark	# of MLBs	Lat. (µs)	Rfig. Data (KB)	MLB En (nJ)	Intcnt. En (pJ)	Op. per MLB	Op. per Cycle
AES	6	0.18	43	6.26	211	6.67	4.44
ATR	12	0.12	67	8.38	1226	6.83	13.67
Census	7	0.10	21	5.81	336	4.57	6.40
CI	9	0.14	17	0.85	664	4.22	5.43
2D-DCT	10	0.38	25	9.93	1504	12.30	6.47
DWT	6	0.12	17	0.90	633	6.83	6.83
MEC	8	0.26	29	5.61	287	4.63	2.85
ME	8	0.18	50	3.15	490	4.25	3.78
SHA-1	12	0.32	87	11.41	1903	11.08	8.31
SW	5	0.56	44	2.69	344	18.80	3.36

Table 17.5 Energy and performance requirement for computation and reconfiguration in MAHA for large input data size

Bench-marks	Data Size (KB)	MAHA Exec En(mJ)	Rfig. En (mJ)	Txfer En (mJ)	MAHA Exec Dly(s)	Rcfig. Dly (ms)	Txfer Dly (s)
AES	256	13.57	0.30	7.03	0.04	0.83	0.12
ATR	256	0.25	0.47	122.03	0.003	0.84	2.14
SHA	609	2.07	0.60	0.01	0.05	0.85	0.001
ME	256	3.81	0.35	0.13	0.02	0.84	0.002
SW	512	6.40	0.30	0.03	1.17	0.83	0.001
DCT2	1024	1.50	0.17	55.37	0.05	0.82	0.97
DWT	1024	0.15	0.12	30.62	0.01	0.82	0.54
MEC	2048	0.51	0.20	0.17	0.02	0.82	0.002
CI	1024	0.18	0.12	90.71	0.02	0.82	1.58
Census	2048	1.29	0.15	0.01	0.02	0.82	0.001
Avg. Contrib(%)		**47.69**	**7.02**	**45.29**	**56.49**	**0.97**	**42.53**

Table 17.6 Improvement in execution time, on-chip and off-chip access count for the baseline CPU with MAHA

Benchmark	Output Data Size(KB)	% Improv. in cycle Cnt	% Improv. in on-chip traffic	% Improv. in off-chip traffic
AES	256	96.8	96.6	3.26
ATR	256	90	87.7	3.16
SHA-1	1	99.9	99.9	9.5.6
ME	256	100	99.9	94.9
SW	1	99.6	99.6	99.7
2D-DCT	1,024	98.2	98.1	75.8
DWT	1,024	20.8	15.4	87.3
MEC	2	99.9	99.8	98.3
CI	1,024	31.8	24.7	56.9
Census	0.1	100	99.9	99.7

17.2.1.1 Reduction in On-Chip and Off-Chip Communication

Table 17.6 lists the improvement in processor execution time, on-chip traffic (L1 and L2 access) and off-chip traffic (L2 miss) compared to the baseline GPP results presented in Table 15.3. From Table 17.6 note that for data-intensive applications such as *DWT* and *Color Interpolation* where the output data size is of similar order of magnitude to the input data size, execution time and energy is dominated by data transfer from off-chip to on-chip memory. For these applications therefore, a lower improvement in off-chip traffic is observed. On an average, *75%* improvement in execution time is observed on the baseline processor, *71%* in on-chip traffic and

17.2 Energy and Performance for Mapped Applications

45% improvement in off-chip traffic. Note that maximum savings is for *Mapreduce* class of applications which have large input and small output data set and are not compute-intensive.

17.2.1.2 Improvement in Execution Latency

The total time to process data in a hybrid (software + hardware) system is calculated as the sum of the reconfiguration time, execution time and the time required to transfer the resultant output to the processor ($T_{total} = T_{reconfig} + T_{exec} + T_{txfer}$). From Tables 17.5 and 17.6, it may be noted that for applications with large output data size, improving the accelerator performance (by increasing frequency or through parallelism) is expected to have minimal impact on the system performance. For the selected benchmarks, a hybrid system with *MAHA* achieves a $8.3\times$ improvement compared to a software only solution.

17.2.1.3 Improvement in Energy

The total energy requirement for the hybrid system is a sum of the reconfiguration energy, the execution energy and the energy to transfer the resultant output to on-chip memory ($E_{total} = E_{reconfig} + E_{exec} + E_{txfer}$). Comparison with the energy required by the software based execution flow (refer to Table 15.3) is shown in Fig. 17.2b. On an average, the off-chip *MAHA* framework improves the total energy requirement for the entire system by $10.9\times$.

17.2.1.4 Improvement in Energy-Efficiency (EDP)

Figure 17.2c compares the energy efficiency for the applications mapped. On an average *MAHA* improves the energy-efficiency by $91.2\times$. The times improvement in energy-efficiency compared to the baseline CPU model varies over a large range, from *1.5X* for *DWT* to *4900X* for *Census* application. This suggests that not all applications are amenable to acceleration with *MAHA* framework. The applications studied are therefore categorized as:

 i. Applications (e.g. *Census*) which are purely data intensive where the output data size is significantly less compared to the input data size benefit most.
 ii. Applications (e.g. *2D-DCT*) which are moderately compute intensive, but the output data size is equivalent to the input data size benefit less.
 iii. Applications (e.g. *AES*) which are compute-heavy and the output size is same or less than input size benefit still less.
 iv. Applications (e.g. *color interpolation*) with low computation requirement and the output size is same as input data size are least likely to benefit.

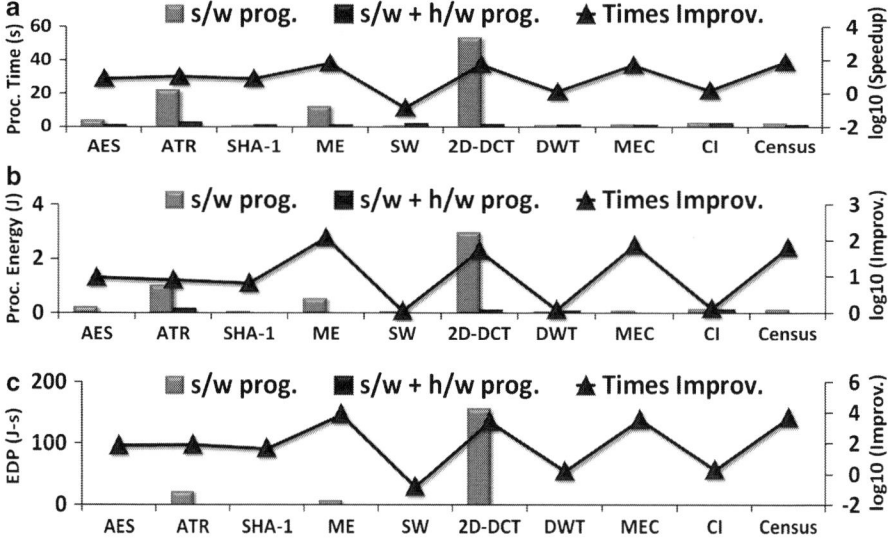

Fig. 17.2 Comparison of (**a**) total processing time; (**b**) total energy requirement and (**c**) energy-efficiency (measured in terms of EDP) between a CPU only system and a system with CPU and MAHA based hardware accelerator

17.2.2 Comparison with FPGA And GPU

Energy required by the off-chip *MAHA* framework has been compared against commercially available FPGA and GPU platforms. In case of FPGA, all benchmarks were mapped to 45nm *Stratix IV* [4] device from *Altera* using Quartus II mapping software. Data from off-chip Flash memory was buffered into multiple 1-port RAM 144k memories on the FPGA. For each benchmark, the energy requirement was calculated based on a dataset with an average activity of 20% using FPGA power simulation tool. For GPU, the benchmarks were run on a platform with an Intel Core i7 2960xm, 32GB of DDR3-1333 memory and 2xGTX 580M, a mobile graphics solution. The GPU is based on an under-clocked GF114 core architecture, realized with 40 nm TSMC process. Benchmarks were developed for this architecture using standard C implementations converted to the CUDA extension. Each benchmark was launched using a program-specific configuration designed to maximize thread occupancy, and with enough blocks to guarantee that all computing elements on the card were active simultaneously. This allowed the average power consumption value for the card itself to be used in the calculations without considering which portion of the card is active. Average GPU power is calculated by multiplying the 40 nm GPU power density, $0.3\,\text{W/mm}^2$, by the die area, $360\,\text{mm}^2$, then reducing this value by 25% to account for the 25% reduction in clock frequency for the mobile version of the GPU. Per-operation energy is calculated by multiplying the average latency per thread by the average GPU power. Figure 17.3 shows a comparison of the energy

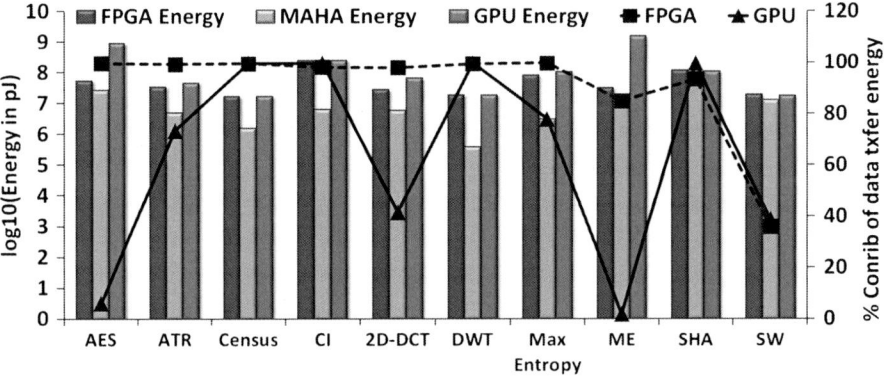

Fig. 17.3 Comparison of energy requirement for FPGA, *MAHA* and GPU frameworks. Contribution of data transfer overhead to overall FPGA and GPU energy requirements is also shown

requirement for FPGA, *MAHA* and GPU frameworks. From Fig. 17.3, it may be noted that on an average *MAHA* improves the energy requirement by *74%* and *84%* over FPGA and GPU frameworks respectively. *MAHA* eliminates the high energy overhead (*91%* and *64%* respectively) for transferring data from off-chip memory to on-chip FPGA or GPU compute engines.

17.3 Hardware Emulation Based Validation

A FPGA-based emulation framework was developed which validates (i) functionality and synchronization of multiple MLBs for several application kernels; and (ii) interfacing the *MAHA* framework with the host processor. The emulation framework attempts to use the off-chip Flash memory simply as a storage without any design modifications to enable in-memory computing. *MAHA* behaves as an energy-efficient loosely-couple off-chip accelerator to which data-intensive kernels can be off-loaded. A similar approach has actually been commercialized [5], in which massively parallel processing is achieved at disk storage interface using FPGAs. By performing in-database processing, latency for analysis workloads is substantially minimized. This section investigates the benefits of replacing FPGA using a more energy-efficient *MBC* model realized through a MAHA framework.

The spatio-temporal *MAHA* computing model with hierarchical interconnect was mapped to a FPGA based hardware emulation framework and this in turn was interfaced with non-volatile solid-state drive (SSD) arrays for off-chip acceleration. The overview of the scheme is illustrated in Fig. 17.4a with architecture and system details provided in Fig. 17.4b. The hardware emulation framework was developed in the Altera Stratix IV FPGA environment using Quartus II and the Nios II soft processor system. The emulation framework consists of two FPGA boards, one

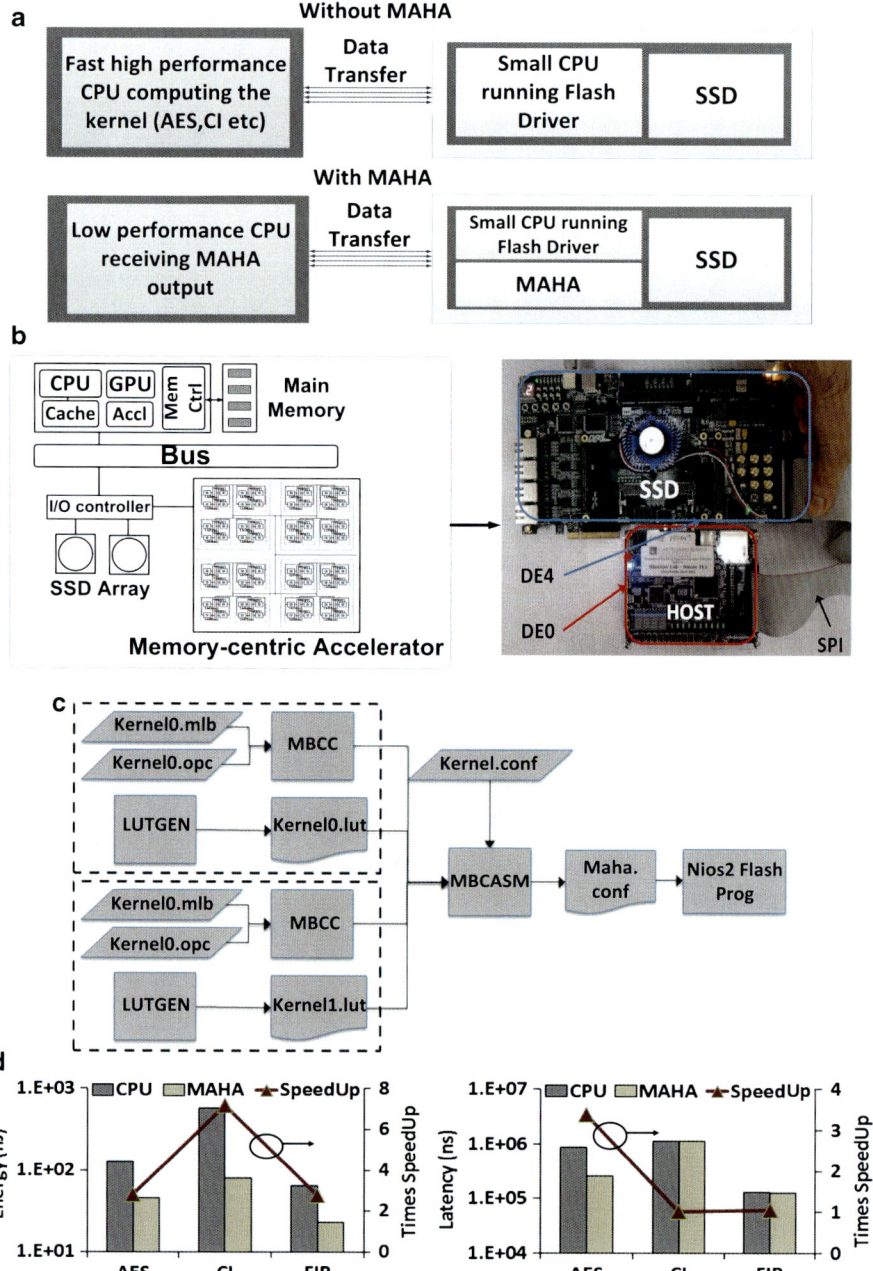

Fig. 17.4 (a) Overview for off-chip acceleration with *MAHA* framework. (b) System architecture for FPGA based hardware emulation framework. (c) Application mapping details for off-chip acceleration. (d) Improvement in latency and energy with *MAHA* based off-chip acceleration

17.3 Hardware Emulation Based Validation

DE0, running a host CPU, and a DE4, consisting of three main components: the *MAHA* framework, the Flash controller, and on board Flash memory. The Flash controller is a light-weight Nios II/e core, and the on board Flash memory has a capacity of 64MB. In comparison, the host CPU is the fastest available Nios II/f processor. The two boards communicate over 3-wire SPI in a simple master/slave configuration. The host CPU sends one 32-bit header to the slave, specifying kernel id (KID) to be run. The slave queries the flash for all available kernels, and, upon finding a match, begins a transfer of the configuration bits and data for processing to the *MAHA* framework. If no match is found, the slave immediately responds with an error code. Otherwise, the slave will only interrupt the host CPU when it has completed the operation, sending the output back to the host. Applications are mapped to this emulation framework using the software flow illustrated in Fig. 17.4c.

Applications are first described in a custom *MAHA* assembly language (kernel.mlb). The language defines common arithmetic and logical operations and also supports a variable number of custom opcodes. These opcodes can be invented as needed by the programmer, only requiring that a second file (kernel.opc) containing a listing of all opcodes used be provided to the compiler (MBCC). The lookup tables are specified in kernel.lut file and are currently generated using a C# script (LUTGEN) where the user can specify any multi-input function. MBCC maps the kernel description to *MAHA* instruction-set architecture (ISA) using the opcode definitions. The code file output from MBCC and the LUT file for each kernel is then picked up by the assembler (MBCASM) and assigned to one or more MLBs based on an assignment information provided in kernel.conf. The assembler generates a binary/object file which contains specific information for each kernel such as start address, length of data, followed by schedule table and function table values for each kernel. This binary file is used to configure the *MAHA* framework using nios-II Flash programmer tool. Functional correctness of the *MAHA* framework was validated by mapping one security kernel (AES) and two signal processing applications (2D color interpolation and FIR filtering). Energy values were obtained using DE4's onboard current sensor and latency from SignalTap Logic Analyzer. The latency and energy values for the SPI component were scaled to reflect the bandwidth and energy values for more realistic PCIe 3.0x16 interconnect on a Stratix IV FPGA platform. Based on cycle-accurate experimental data (Fig. 17.4d), it may be noted that 4X speedup in latency is observed for AES while 3X, 7X and 3X improvement in energy requirement was achieved for the three benchmarks. EDP improvements of the order of 10X, 7X and 3X was achieved on the emulation platform. Note that the benefit is due to off-chip computing in proximity to the last level of memory.

References

1. V. Mohan, S. Gurumurthi, M.R. Stan, "FlashPower: A Detailed Power Model for NAND Flash Memory", in *DATE*, 2010
2. [Online]. "Micron 1Gb NAND Flash Data Sheet". http://download.micron.com/pdf/datasheets/flash/nand/1gb_nand_m48a.pdf
3. "Simplescalar Toolset v3.0" http://www.simplescalar.com/
4. "Stratix V Device Handbook" http://www.altera.com/literature/hb/stratix-v/stx5_5v1.pdf
5. "Netezza Data Warehouse Appliances" http://en.wikipedia.org/wiki/Netezza

Part VII
Improving Reliability of Operations in MBC

Chapter 18
Mitigating the Effect of Parametric Failures in MBC Frameworks

Abstract Since MBC uses large, high-density memories for computation, reliable operation of the framework under increasing process variations becomes a major concern. Variation may potentially cause memory access failures or flipping of stored data during read-out (Mukhopadhyay et al., IEEE Trans Comput Aided Des Integrated Circ Syst 24(12), 2005), which leads to incorrect execution of the mapped application. Moreover, in order to reduce the power requirement, memory core is conventionally operated at lower supply voltages. Although this minimizes the active and leakage power consumption, read and access failure probabilities increase significantly at low operating voltages (Mukhopadhyay et al., IEEE Trans Comput Aided Des Integrated Circ Syst 24(12), 2005). In order to address variation induced memory failures in the MBC framework, it's read dominant behavior can be exploited since write to the MBC framework only occurs occasionally during reconfiguration. Exploiting the read-dominant memory access pattern in MBC, a preferential memory cell sizing approach is proposed to make the memory more robust to read and access failures. The resultant decrease in write stability is addressed by low-overhead circuit techniques during reconfiguration.

18.1 Effect of Parameter Variations on QoS

Parametric variation in nanoscale technologies can lead to incorrect execution of applications mapped to the MBC framework. Such errors in execution may arise from any one of the following sources: (i) *Failure in memory*, (ii) *Failure in Control Unit* and (iii) *Failure in programmable interconnects*. Memory failures due to degraded cell stability is most difficult to address among all the parametric failures. Such failures can eventually lead to Quality of Service (QoS) degradation at the MBC output.

Variation in process parameters (both systematic and random) have emerged as a major design challenge at both circuit and architecture level design [1]. Such variations can cause mismatch between adjacent transistors in a SRAM cell, which may eventually lead to a failure of the cell itself. These failures include: (i) *Access Failure* which results due to an increase in the cell access time; (ii) *Read Failure* due to an unwanted flipping of the cell content while reading; (iii) *Write Failure* due to the inability to successfully write to the memory cell and (iv) *Hold Failure* which results due to the application of a supply voltage lower than the minimum voltage required by the cell to hold the data at nominal condition. Since, the above failures are caused by variation in device parameters, they are often referred to as parametric failures [2]. Although alternate failure mechanisms exist for memory cells [2], the focus is on the effect of parametric failures on MBC and their mitigation.

Let us consider a memory model consisting of 16KB of memory, divided into 128 blocks, each with 1,024 cells. The memory cells in each block is organized into 32 rows (N_{ROW}) with 32 cells (N_{COL}) in each column. Note that depending on the maximum number of partition inputs and outputs specified during partitioning, a single memory block can store the entire LUT or only a part of it. For example, for mapping a LUT with 12 address bits and 8 data bits, 32 similar blocks will be necessary. The appropriate 8 bit selection will be performed through a combination of row-decoding and column-multiplexing. The number of redundant columns in each block (N_{RC}) is 2. A pulsed wordline and bitline isolation architecture is also considered for reducing bitline swing and a nominal V_{dd} of 0.9 V. Monte Carlo simulations were performed with this memory model using HSPICE for PTM 45nm LP models [3] considering both systematic and random variation in V_T. Details of the simulation can be obtained from [4].

In the Monte Carlo (MC) simulations performed on the cells inside each block, the parameters V_{read}, V_{trip}, T_{access} and T_{write} were observed for each cell. These parameters defined in [4] are commonly used to characterize the vulnerability of memory cells to parametric failures [2]. As pointed out in [5, 6], a "weak" memory cell can undergo a read/write or access failure under supply voltage variation, high temperature or coupling noise. It is therefore necessary to identify such "weak" cells through post-silicon calibration techniques. In order to quantify the vulnerability of individual memory blocks, a block level reliability metric is introduced, obtained by collating V_{read}, V_{trip}, T_{access} and T_{write} parameters from individual memory cells. The metrics are:

- Indicator for Read Stability denoted as $I(V_{trip} - V_{read})$
- Indicator for Write-ability denoted as $I(T_{write})$
- Indicator for Access-ability denoted as $I(T_{access})$

In general the indicator $I(x)$ where $x \equiv V_{trip} - V_{read}$, T_{write} or T_{access} for a memory block can be derived by:

- Classifying the cells of the block into separate bins based on the value of parameter x
- Calculating a weighted average of the cells in each bin:

$$I(x) = \frac{\sum_{i=1}^{N} n_i * w_i}{\sum_{i=1}^{N} w_i} \quad (18.1)$$

18.1 Effect of Parameter Variations on QoS

In (18.1), w_i denotes the weight and n_i denotes the number of cells in the ith bin. For the simulations, the cells in each block were divided into 5 bins ($N = 5$) with $w_i = 2^{-i}$ values. The range of $x \equiv V_{trip} - V_{read}$ values for the bins are : (i) $x \leq 0$, (ii) $0 < x \leq 100\,\text{mV}$, (iii) $100\,\text{mV} < x \leq 150\,\text{mV}$, (iv) $150\,\text{mV} < x \leq 200\,\text{mV}$, and (v) $x > 200\,\text{mV}$. Choice for the number of bins and the range of x values were primarily driven by the distribution of the $V_{trip} - V_{read}$ obtained from the simulations. For low supply voltages, considering a parametric variation scenario with $\sigma_{dVt_{sys}} = \sigma_{dVt0_{rand}} = 50\,\text{mV}$, $V_{trip} - V_{read}$ values for the "weak" cells were mostly centered around a value of 50 mV. Hence the 2nd bin was chosen for $0 < x \leq 100\,\text{mV}$. The weight assigned to each bin is primarily determined by the operating condition of the MBC framework and the relative vulnerability to failure for cells in each bin. Ideally the weights for the "weak" cells should be exponentially large in order to distinguish them from cells with $x > 100\,\text{mV}$.

$I(x)$ for a given block serves as an indicator of the process corner to which the block has moved due to systematic variation. A block with a larger value of $I(V_{trip} - V_{read})$ is more prone to read disturb failures under increased environmental stress (i.e. reduced voltage and or higher temperature). Figure 18.1a shows the inter-block distribution of $I(V_{trip} - V_{read})$ at the fast corner.

In order to observe the effects of variation induced failures in the memory on the performance of the MBC framework, a discrete cosine transform (DCT) was mapped to the MBC framework. Details of the mapping procedure is described in [4]. LUTs for the DCT operation were first randomly mapped to the blocks of the memory model. Relative vulnerability to parametric failures for these blocks were determined through Monte Carlo simulations considering the systematic and random intra-die components mentioned above. For the distribution of process parameters obtained after Monte Carlo simulations, a reliability map of the memory blocks is generated on the basis of their $I(V_{trip} - V_{read})$ values. This reliability map serves as an input to the variability-aware preferential mapping approach. Note that:

- At lower supply voltages, under parametric variation, the failure probability for the SRAM cell increases significantly [2]. However, the relative vulnerability of the blocks due to systematic variation remains unchanged.
- Due to its input dependence, an average of the PSNR at the DCT output over a length of time fails to capture the worst-case scenario. A worst-case PSNR for multiple input vectors is therefore a better metric than a simple *time average*. However, due to the probabilistic nature of the parametric failures, estimation of true worst-case degradation is infeasible. An average over multiple instances of the same memory design yields the average worst-case PSNR ($Avg_{PSNRwrst}$) which serves as a comparison metric in the QoS analysis.

Figure 18.2 shows the output image quality ($Avg_{PSNRwrst}$) for random mapping of the LUTs to the memory blocks. Random mapping means that the critical and non-critical computations of the target applications are mapped to memory blocks without considering their relative vulnerability to parametric failures. The final

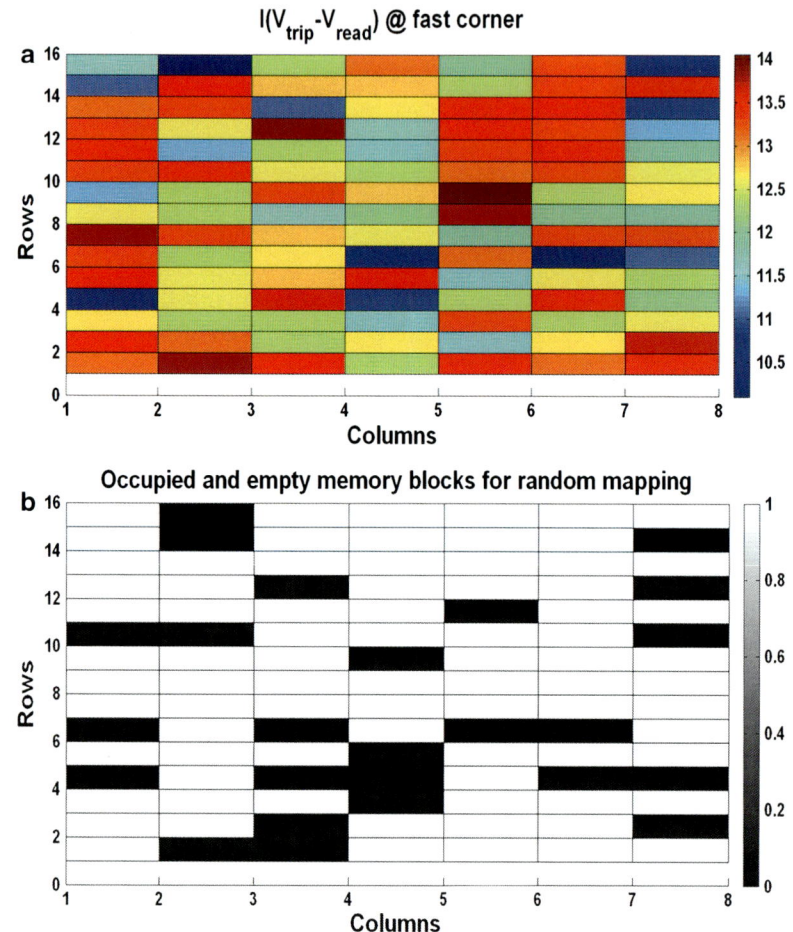

Fig. 18.1 (**a**) shows the variation in $I(V_{trip} - V_{read})$ for the memory blocks at fast corner. (**b**) Random mapping of DCT modules to memory blocks

image is obtained by taking IDCT of the DCT components obtained as outputs from the MBC framework. Under variation, due to read failures in the LUTs, these DCT components will be different from the values obtained from the MBC framework in a no-variation scenario. As shown in Fig. 18.2b, due to random assignment, the final image quality suffers considerable degradation with an average worst-case PSNR of 28.49 dB.

To alleviate degradation in output quality due to variation induced memory failures in the MBC framework, a three-step solution is proposed that involves joint circuit/architecture level optimizations. As demonstrated later, the proposed approach can provide significant improvement in parametric yield under variations.

18.2 A Variation-Aware Preferential Mapping Approach

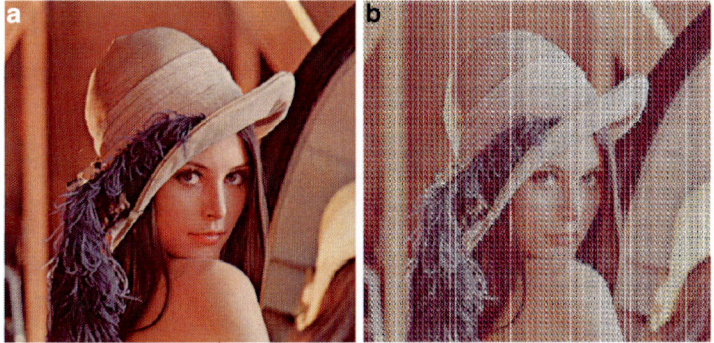

Fig. 18.2 (**a**) Original image at input of DCT; (**b**) Image after inverse DCT, where 1D-DCT operation is performed using the MBC framework. ($Avg_{PSNRwrst} = 28.49$ dB)

The approach can also be used for power reduction using aggressive voltage scaling. As summarized in Fig. 18.3, the major steps in the proposed co-design approach are:

- A preferential (skewed) design of memory cell using transistor sizing that increases the read stability at the expense of write failures (which can be compensated during occasional reconfiguration using several low-cost techniques such as lower cell supply during write) for the read-dominated MBC framework.
- Post-fabrication characterization of the memory and generation of the reliability map to store $I(V_{trip} - V_{read})$, $I(T_{write})$ and $I(T_{access})$ values for the memory blocks.
- A preferential application mapping approach that maps the critical computations to more reliable memory blocks under delay constraint.

18.2 A Variation-Aware Preferential Mapping Approach

Many signal processing applications (such as DCT, FIR filtering) consist of critical and less critical computations [7]. A mapping algorithm which: (a) partitions the computations into critical and less-critical bins; and then (b) maps the critical computations to more reliable sections of the memory can achieve a substantial performance improvement over a random mapping approach.

18.2.1 Post-Si Generation of Reliability Map

A number of techniques [5, 6, 8] have already been proposed to generate post-Si reliability map for embedded memories in order to cope with device variation induced parametric yield loss. These techniques either use March Test [9] to detect

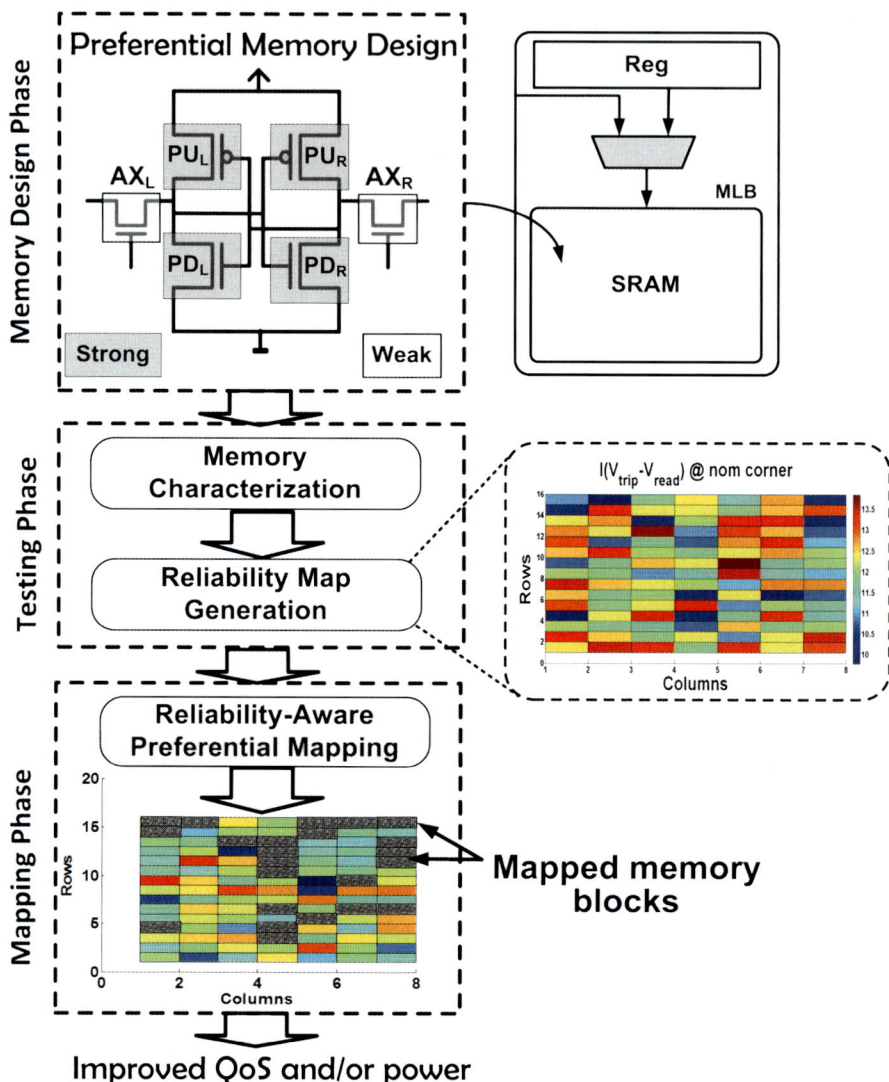

Fig. 18.3 Methodology for robust and low-power operation in MBC

parametric failures or attempt to directly measure the read/write margin for memory cells [6]. The trade-offs between these two alternative approaches are provided in [4]. Since the motivation for reliability map generation is only to identify the relative variability of the memory blocks, an array based technique can be used for fast characterization of the memory blocks. With array-based test approaches, it is possible to determine the relative variability of the memory blocks on the same die. The reliability map used for preferential application mapping is stored

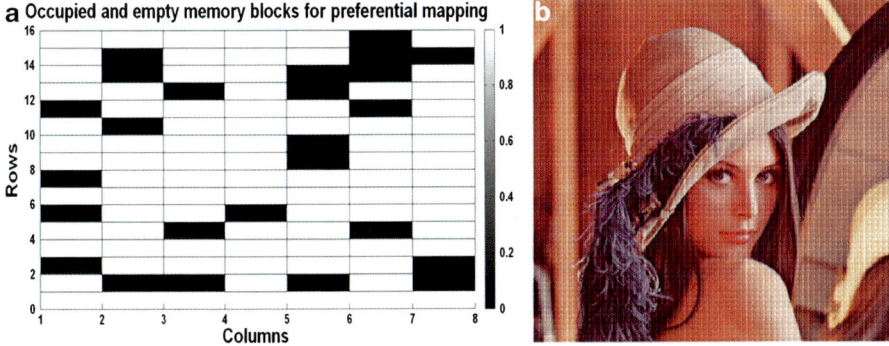

Fig. 18.4 (a) Preferential mapping of DCT modules to memory blocks (b) Image after inverse DCT, where 1D-DCT operation is performed using the MBC framework. ($Avg_{PSNR_{wrst}} = 32.48$ dB)

on a non-volatile memory which is loaded with the relative reliability values of the memory blocks during manufacturing test. If an array based characterization is pursued for map generation with block size of 1KB, the total memory requirement to store the reliability map for a function table of size 16KB is only 64 bits. This follows from the observation that the $I(V_{trip} - V_{read})$ values can be encoded into 4 bits without significant quantization error (refer to Fig. 18.1a).

18.2.2 Impact of Preferential Mapping on Output Quality

A heuristic-based preferential mapping approach can substantially improve the output quality of the target application under parameter variation. If $c_i (i=1 \cdot \cdot P)$ denotes the contribution of the ith computation to the overall output quality and $r_j (j = 1 \cdot \cdot N, N \geq P)$ denotes the reliability measure of the jth memory block, then the mapping heuristic should attempt to maximize $\sum_{i=1}^{P} c_i r_j$. The reliability measure for the jth memory block is represented by $r_j = \frac{1}{I_j(x)}$. A simple minded approach is to sort $C = \{c_i\}$ and $R = \{r_j\}$ in descending order of their values, and then to assign the computations in the order of their contribution to unassigned memory blocks with highest value of r_j.

As an example, in the DCT application, first adder stage and the precomputer blocks are assigned to minimum $I(V_{trip} - V_{read})$ blocks for more reliable operation. Since probability of read failure (P_{RF}) dominates over the other parametric failure mechanisms at low frequencies of operation, assignment of blocks have been based upon $I(V_{trip} - V_{read})$. However, the same can be achieved based on $I(T_{access})$, $I(T_{write})$ or on a weighted combination of all. Figure 18.4a shows the memory block assignment using the preferential mapping approach. The unassigned memory blocks are highlighted in black. A comparison with Fig. 18.1a clearly shows that blocks with high $I(V_{trip} - V_{read})$ have been avoided through the proposed mapping approach. The output as shown in Fig. 18.4b achieves about 4dB improvement in PSNR over a random mapping policy.

18.2.3 Preferential Mapping Algorithm Under Delay Constraint

A preferential mapping approach which does not consider any delay constraint would attempt to provide the best output QoS for a given variation model. However, this may cause increased delay overhead due to larger spacing between the computing elements in the critical path. Thus, in order to achieve the optimal mapping (in terms of reliability) under a delay constraint, resource allocation should be a part of the placement and routing step. A heuristic-based post-processing of the delay optimal routed netlist has therefore been developed that maximizes QoS without violating the delay constraint. Figure 18.5 shows the timing-driven variability-aware preferential mapping process. It starts with ranking the computations in order of their criticality and the memory blocks in order of their reliability measures (defined as r_i for the ith memory block). In descending order of criticality, each computation in the routed design is then considered for reassignment if it is mapped to a memory block with unacceptable r_i. The reassignment step first looks for an unused block with higher r_i and at the smallest Euclidean distance d_i from its current position. The impact on timing is calculated for each trial using incremental timing analysis. If the resulting placement exceeds the delay constraint, the heuristic attempts to swap the current computation with a less critical one which is assigned to a more reliable memory block. As before, computing elements at the smallest Euclidean distance are given higher priority and timing for the new placement is checked so that it does not violate the delay constraint. The major steps of the heuristic is illustrated in Fig. 18.5.

Reliability map for the entire memory array as obtained from Monte Carlo simulations is shown in Fig. 18.6b. With this reliability distribution, PSNR degradation at the DCT output was measured for the preferential mapping approach. Figure 18.6a shows the original routed design from VPR and the output from the timing-driven preferential mapping heuristic presented in Fig. 18.5c. The modified placement as shown in Fig. 18.6c has the same delay (4.78 ns) as the delay-optimal placement shown in Fig. 18.6a. However, the output PSNR improves from a value of *28.49* dB to *31.72* dB.

18.2 A Variation-Aware Preferential Mapping Approach

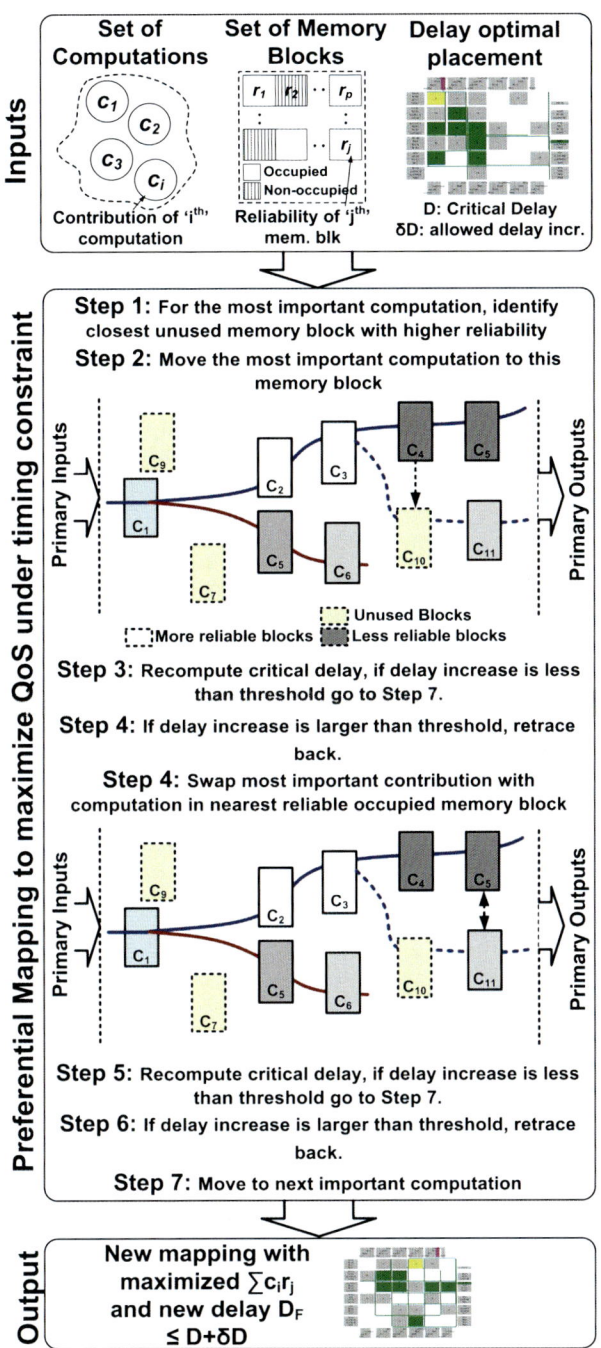

Fig. 18.5 Flowchart showing the major steps for the heuristic-based preferential mapping approach under timing constraint

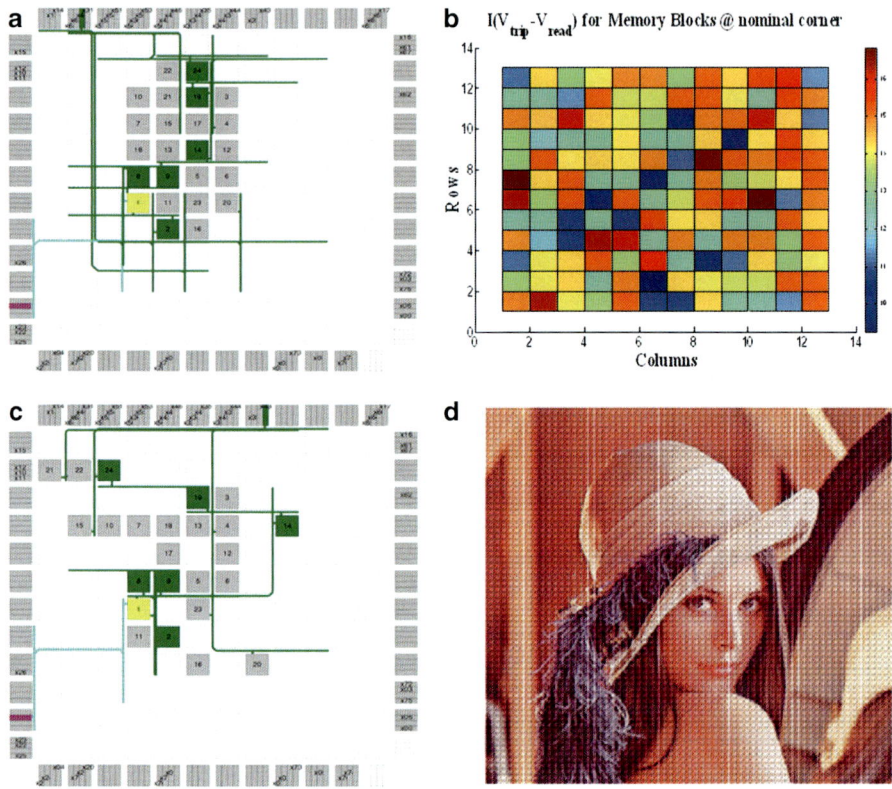

Fig. 18.6 (**a**) Delay optimal placement of the DCT blocks obtained from VPR [10]; (**b**) $I(V_{trip} - V_{read})$ for the memory based computing elements used to map the DCT application; (**c**) Final placement after the preferential mapping under timing constraint; (**d**) Final image after mapping has a PSNR of 31.72 dB

18.3 Preferential Memory Design

In a SRAM cell design, optimization of read and write stability present contradictory requirements. However, the read-dominated access pattern in the MBC framework can be leveraged to trade off read stability with write-ability of memory cells. Stability of occasional write operation in MBC during reconfiguration can be improved by dynamic change in operating conditions. Although a number of novel memory schemes have been proposed to improve the read stability of the SRAM cell [11, 12], this work explores a sizing based approach which incurs negligible design overhead.

18.3 Preferential Memory Design

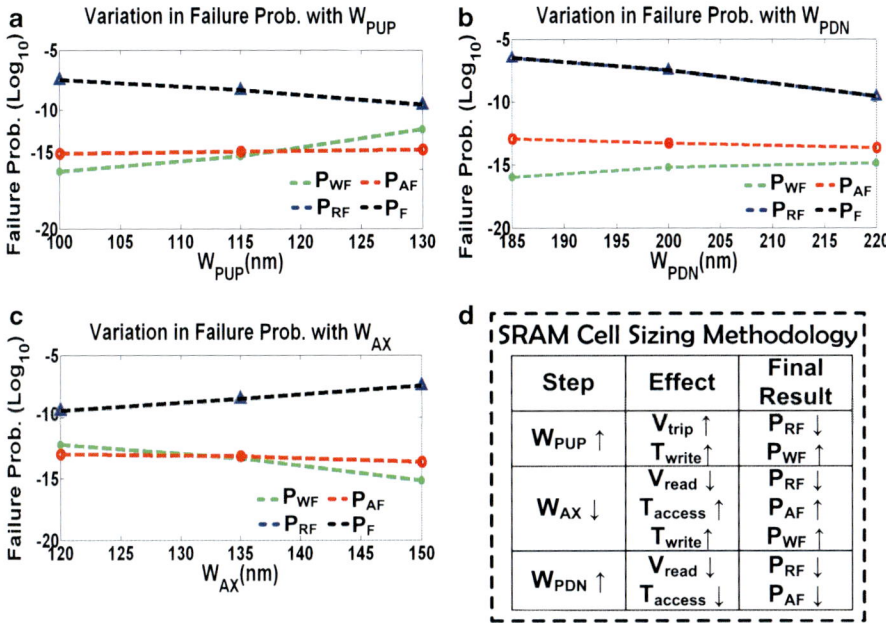

Fig. 18.7 Variation in failure probabilities with (**a**) W_{PUP}, (**b**) W_{PDN} and (**c**) W_{AX}; (**d**) SRAM cell sizing methodology

18.3.1 Preferential Sizing of 6-T SRAM

The criteria followed for preferential sizing of the 6-T SRAM cell are: (i) Minimize the read and access failure probabilities; (ii) Minimize the increase in area of the memory cell. The cell area estimation methodology is the same as followed in [2]. Figure 18.7 shows the dependence of the read, write and access failure probabilities on the width of the pull up (W_{PUP}), pull down (W_{PDN}) and access (W_{AX}) transistors. Details of calculation for the various probabilities are provided in [4]. From Fig. 18.7a–c it may be noted that:

- Increasing W_{PUP} increases V_{trip} and reduces P_{RF}.
- Increasing W_{PDN} decreases V_{read} and thus reduces P_{RF}.
- Decreasing W_{AX} reduces P_{RF}.
- Increase in P_{AF} due to weakening of W_{AX} can be compensated by increasing W_{PDN}.
- First two optimizations increase T_{write} and P_{WF}.

The nominal memory cell was skewed based on the above observations following the methodology as outlined in Fig. 18.7d. The old and new value for the transistor sizes are given in Table 18.1. With $L_{min} = 45$ nm, the increase in area per cell was calculated to be 2.5% according to the formula provided in [2]. Figure 18.8a–c show the distributions of V_{read}, T_{write} and T_{access} for the nominal and the skewed cell.

Table 18.1 Transistor widths for nominal and skewed memory cell

Cell	W_{PUP}(nm)	W_{AX}(nm)	W_{PDN}(nm)
Nominal	100	150	200
Skewed	125	125	220

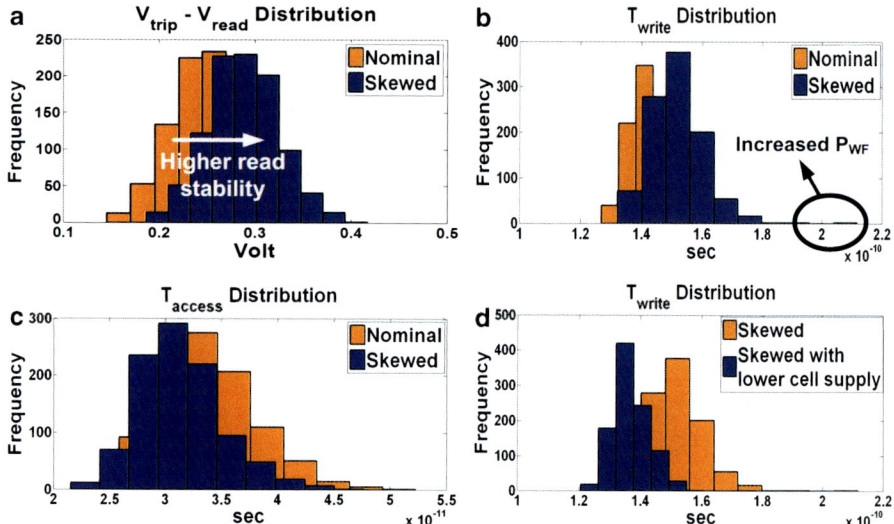

Fig. 18.8 Distribution of: (**a**) $V_{read} - V_{trip}$, (**b**) T_{write} and (**c**) T_{access} for the nominal and the skewed memory cell. (**d**) Shows the T_{write} distribution for cell with column-based lowering of cell supply voltage

18.3.2 Compensation for Increased Write Failure Probability

From Fig. 18.8a and c, it may be noted that the skewed memory design achieves better tolerance to read and access failures at the cost of higher write failures (refer to Fig. 18.8b). Since a higher wordline voltage degrades the half-select stability of the cells in the same row [13], a column based lowering of cell supply is proposed to achieve better write-ability of the skewed memory cell. Note that such column based dynamic control of the supply voltage has been explored in earlier works [14] to improve both read and write stability while minimizing leakage power. Although such a scheme incurs area overhead due to complicated power grid and routing, it achieves nearly a 10X reduction in single bit fails by creating a 100 mV voltage differential between the wordline and cell supply. In the simulation framework, supply voltage for the memory cell was reduced to 0.7 V to improve T_{write} (Fig. 18.8d). Table 18.2 shows the overall failure rate for the nominal design, the skewed design as well as skewed design with lower cell supply voltage for write. A skewed design with lower cell supply for write reduces the total cell failure probability (P_F) by factor of 111.94.

18.3 Preferential Memory Design

Table 18.2 Cell failure probabilities for $V_{ddnom} = 0.9$ V

Cell	P_{RF}	P_{WF}	P_{AF}	P_F
Nominal	3.28e−8	6.66e−16	2.92e−14	3.28e−8
Skewed	2.93e−10	2.77e−12	2.22e−14	2.95e−10
Skewed w/ lower WR Vdd	2.93e−10	3.33e−16	2.22e−14	2.93e−10

18.3.3 Impact of Cell Sizing on Output Quality

Figure 18.9a illustrates the fact that skewing leads to redistribution of the cells in the 5 bins for a given memory block. Since a heavy penalty is associated with each failing cell, skewing leads to an improvement of $I(V_{trip} - V_{read})$ values across all the blocks in the memory (Fig. 18.9b). Figure 18.9c–g, illustrates the PSNR improvement achieved through the preferential mapping and memory design. Following points may be noted from Fig. 18.9c–g.

- Fast and slow corners in Fig. 18.9d and e correspond to global fast and slow corners, i.e. for both NMOS and PMOS. A PSNR value of 40dB is used to represent a *no-degradation* case, although the PSNR is theoretically ∞ for no degradation.
- Preferential mapping for the original memory can achieve a significant improvement in PSNR for a range of $I(V_{trip} - V_{read})$ as high as 1.75. Considering that a cell moving from bin #1 to bin #2 reduces $I(V_{trip} - V_{read})$ by 0.25, this is equivalent to tolerating 5 more read failures in the memory block without increasing N_{RC}.
- Due to significantly smaller values of $I(V_{trip} - V_{read})$, the simulations did not exhibit any degradation in output quality for the skewed design. However, if the tolerance for $I(V_{trip} - V_{read})_{max}$ is reduced to smaller value (in other words considering a higher variation for the skewed design), a degradation of the PSNR values is observed at the output (Fig. 18.9c–g). Preferential mapping can again be applied to the latter case to improve performance.
- In the fast-N slow-P corner, the memory cell becomes most susceptible to read failures. This is indicated by a higher value of $I(V_{trip} - V_{read})$ which the system must tolerate to produce an acceptable PSNR at the output.
- Improvement in output PSNR due to preferential mapping after skewing is comparatively smaller than the mapping applied to the original design. The reason for this is evident from Fig. 18.9h which shows that skewing leads to large improvement in read stability for all memory cells, thereby reducing improvement with the mapping approach. The skewed memory cells however occupy larger area compared to the nominal design and experience increased probability of write failures. The preferential mapping approach therefore can provide either a software-only or a complementary solution to circuit-level optimization during design to tolerate the effect of parameter variations.

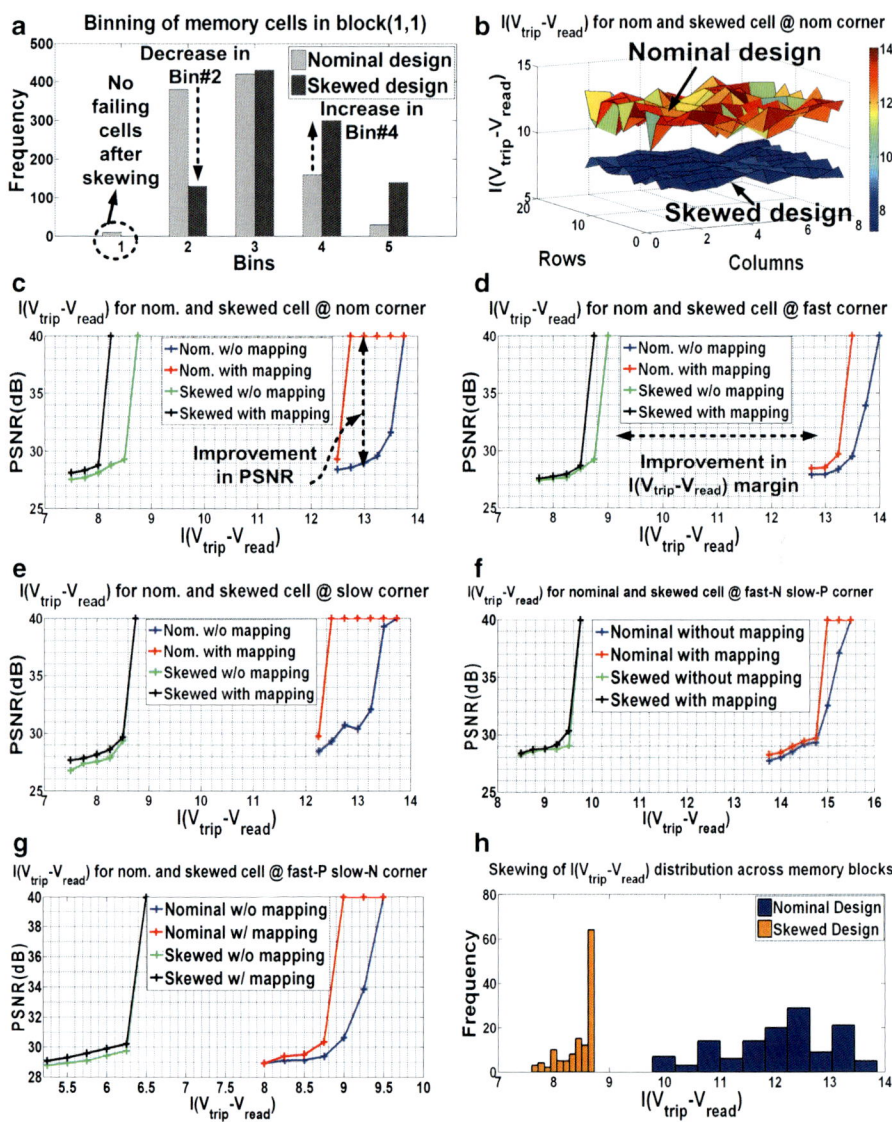

Fig. 18.9 (a) Redistribution of memory cells into bins before and after skewing; (b) Skewed design achieves better $I(V_{trip} - V_{read})$ for all blocks; PSNR values for (i) original without mapping, (ii) original with mapping, (iii) skewed without mapping and (iv) skewed with mapping for (c) fast, (d) nominal, (e) slow, (f) fast-N slow-P and (g) fast-P slow-N corners. (h) Most blocks have moved to a similar $I(V_{trip} - V_{read})$ region after skewing

References

1. S. Borkar, "Microarchitecture and Design Challenges for Gigascale Integration", in *Intl. Symposium on Microarchitecture*, 2004
2. S. Mukhopadhyay,H. Mahmoodi, K. Roy, "Modeling of failure probability and statistical design of SRAM array for yield enhancement in nanoscaled CMOS". IEEE Trans. Comput. Aided Des. Integrat. Circ. Syst. **24**(12), 1859–1880 (2005)
3. [Online], "Predictive Technology Model". http://ptm.asu.edu/
4. S. Paul, S. Mukhopadhyay, S. Bhunia, "A variation-aware preferential design approach for memory based reconfigurable computing". *IEEEE/ACM International Conference on Computer-Aided Design Digest of Technical Papers*, 2009
5. D.M. Kwai et al., "Detection of SRAM Cell Stability by Lowering Array Supply Voltage", in *Asian Test Symposium*, 2000
6. A. Pavlov et al., "Weak cell detection in deep-submicron SRAMs: A programmable detection technique". J. Solid State Circ. **41**(10), 2334–2343 (2006)
7. N. Banerjee, G. Karakonstantis, K. Roy, "Process variation tolerant low power DCT architecture". *Design Automation and Test in Europe*, 2007
8. N.N. Mojumder, S. Mukhopadhyay, J.J. Kim, C.T. Chuang, K. Roy, "Design and Analysis of a Self-Repairing SRAM with On-Chip Monitor and Compensation Circuitry", in *VLSI Test Symposium*, 2008
9. M. Bushnell, V. Agarwal, *"Essentials of Electronic Testing for Digital, Memory and Mixed-Signal VLSI Circuits"* (Springer, Heidelberg, 2000)
10. [Online], "VPR and T-VPack 5.0.2 Full CAD Flow for Heterogeneous FPGAs". http://www.eecg.utoronto.ca/vpr/
11. J.J. Kim, K. Kim, C.T. Chuang, "Asymmetrical SRAM Cells with Enhanced Read and Write Margins", in *International Symposium on VLSI Technology Systems and Applications*, 2007
12. L. Chang et al., "Stable SRAM Cell Design for the 32nm Node and Beyond", in *Symp. on VLSI technology*, 2005
13. J.J. Kim, S. Mukhopadhyay, R. Rao, C.T. Chuang, "Capacitive Coupling Based Transient Negative Bit-line Voltage (Tran-NBL) Scheme for Improving Write-ability of SRAM Design in Nanometer Technologies", in *International Symposium on Circuits and Systems*, 2008
14. K. Zhang et al., "A 3-GHz 70Mb SRAM in 65nm CMOS Technology with Integrated Column-Based Dynamic Power Supply", in *International Solid-State Circuits Conference*, 2005

Chapter 19
Mitigating the Effect of Runtime-Failures in MBC Frameworks

Abstract In nanoscale technologies, memories are vulnerable to both parametric failures, as well as, runtime failures induced by "soft errors" such as voltage, or thermal noise and aging effects. This chapter addresses the runtime failures in on-chip memories induced by "soft errors". Conventionally such runtime errors are addressed using single error correction double error detection (SECDED) codes. However, these codes have very limited correction capability, making them inefficient to protect memory in scaled technologies (sub-45 nm), which are particularly vulnerable to multiple-bit failures (Hareland et al. Characterization of Multi-bit Soft Error events in advanced SRAMs, Intl. Electron Devices Meeting, 2003; Osada et al. IEEE J Solid State Circ 39(5), 2004). The requirement to tolerate multi-bit failures is accentuated by inter-die and intra-die variation in memory blocks which increases the vulnerability towards runtime failures. This chapter explores architectural modifications and a novel reconfigurable error-control coding (ECC) technique which together can be extremely effective in protecting on-chip memories against runtime failures.

19.1 Impact of Parameter Variation on Runtime Failures

Similar to off-chip non-volatile memory arrays, data arrays for on-chip memories such as on-chip caches can be instrumented to realize a reconfigurable framework (MAHA) for on-demand computing. It is therefore essential to protect the on-chip cache against runtime failures. Runtime failures in on-chip cache memories can be classified as: (a) random failures due to voltage/thermal noise and (b) contiguous failures due to soft error. Although the Soft Error Rate (SER) of an SRAM cell is found to remain constant across process technologies, due to the increase in the memory size per generation of the chip, the total number of soft errors increases with processor generation [1, 2]. Moreover, as the SRAM footprint decreases with process scaling, the probability of MBUs due to a single particle strike increases

[1, 3]. Runtime failures in processor caches have been conventionally addressed by ECC. The level of ECC protection is typically different for L1 and L2/L3 caches due to their different design criteria.

Due to process parameter variations, memory cells which are marginally functional during manufacturing test can undergo runtime failures due to voltage or thermal noise, soft-error or aging effects. These cells (referred to as "weak" cells [4, 5] must therefore be protected using more ECC compared to cells which have suffered little variations. As argued in [6], memory cells which have either moved to the low threshold (V_t) or the high V_t region due to inter/intra-die variations are susceptible to read, write, access and hold failures. A similar trend was observed in the simulations presented in [7].

Based on the above analysis, it was noted that a $\pm 10\% \Delta V_t$ variation from the nominal value is sufficient to cause the cell to be more vulnerable to one or more modes of runtime failures. As a result a uniform ECC allocation performed statically for all memory blocks during design time fails to account for the distribution of vulnerability across memory blocks after fabrication. Moreover, it can lead to overly pessimistic results if the worst-case vulnerability of a memory block is accounted for during ECC allocation. In this chapter, a reliability-driven ECC allocation scheme is therefore proposed that matches the relative vulnerability of a memory block (determined using post-fabrication characterization) with appropriate ECC protection. As shown in Fig. 19.1a, the entire spectrum for PMOS and NMOS V_t variation in a SRAM cell can be divided into 3 regions based on the extent of ECC protection they should receive. From this, it may be noted that a strong PMOS and a nominal NMOS ($\Delta V_t < \pm 10\%$ of the nominal value) is most resilient to the failure mechanisms considered above. Cells with such configuration should therefore be assigned the least ECC protection ($t = t_{min}$ where t denotes the number of errors that may be corrected for a given block). Cells with PMOS and NMOS V_t variation less than $\pm 10\%$ of the nominal value should be assigned a higher ECC protection ($t = t_{med}$) due to their vulnerability towards soft error. Cells falling outside the above ΔV_t regions should be assigned the maximum ECC protection ($t = t_{max}$) to tolerate runtime failures arising from system noise, soft errors and aging effects.

19.2 Reliability-Driven ECC Allocation for Multiple Bit Error Resilience

To address the variation in vulnerability to runtime failures based on process variation, a reliability-driven error protection scheme is proposed. The scheme offers tolerance to single and multi-bit failures for different sections of the memory array based on the extent of process variation experienced by the memory array. It combines both architectural modifications as well as novel reconfigurable BCH code implementation.

19.2 Reliability-Driven ECC Allocation for Multiple Bit Error Resilience

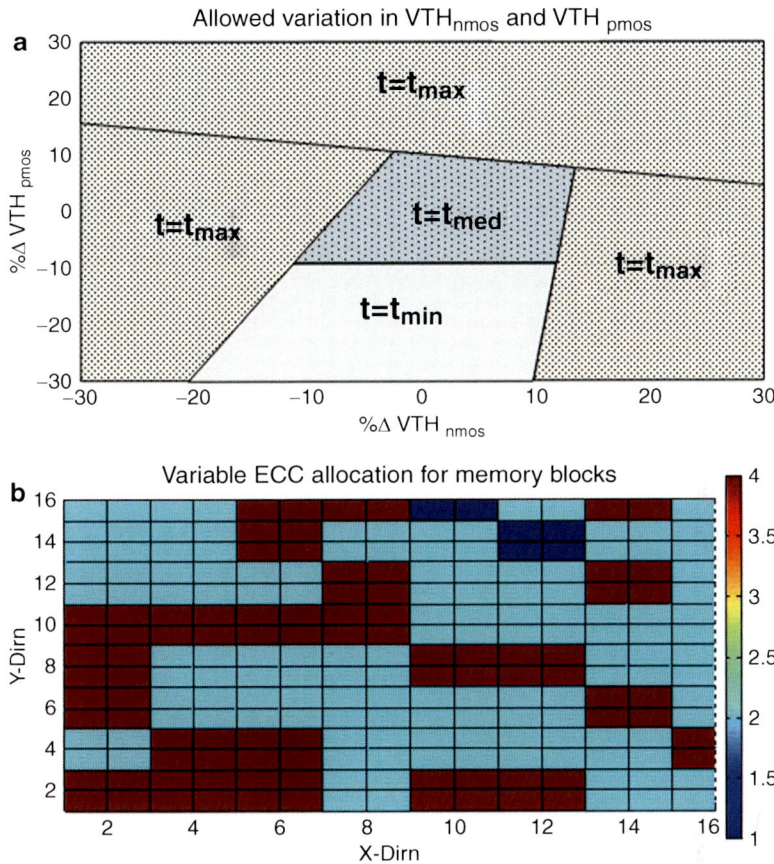

Fig. 19.1 (a) Based on variability, the spectrum of NMOS and PMOS variation is divided into multiple ECC regions. (b) ECC allocation for memory blocks on a die at worst-case inter-die corner

19.2.1 Overall Scheme

Figure 19.2a shows the major steps of the proposed variability aware ECC allocation methodology in large caches. As seen from Fig. 19.2a, the cache size (S), cache associativity (A), block size (B) and the maximum number of bit-errors that can be tolerated (t_{max}) serve as input specifications for the variable ECC allocation methodology. During the design phase, based on t_{max}, the number of required check bits (m) and the number of ways (w) required to store m check bits is calculated. Suitable μ-arch modifications to interchangeably use these "w" ways for data and ECC storage are then incorporated into the cache. Finally, a low-latency and energy-efficient BCH encoder/decoder is implemented at the memory interface to detect and correct up to t_{max} bit-errors. Post-fabrication during manufacturing test, a reliability

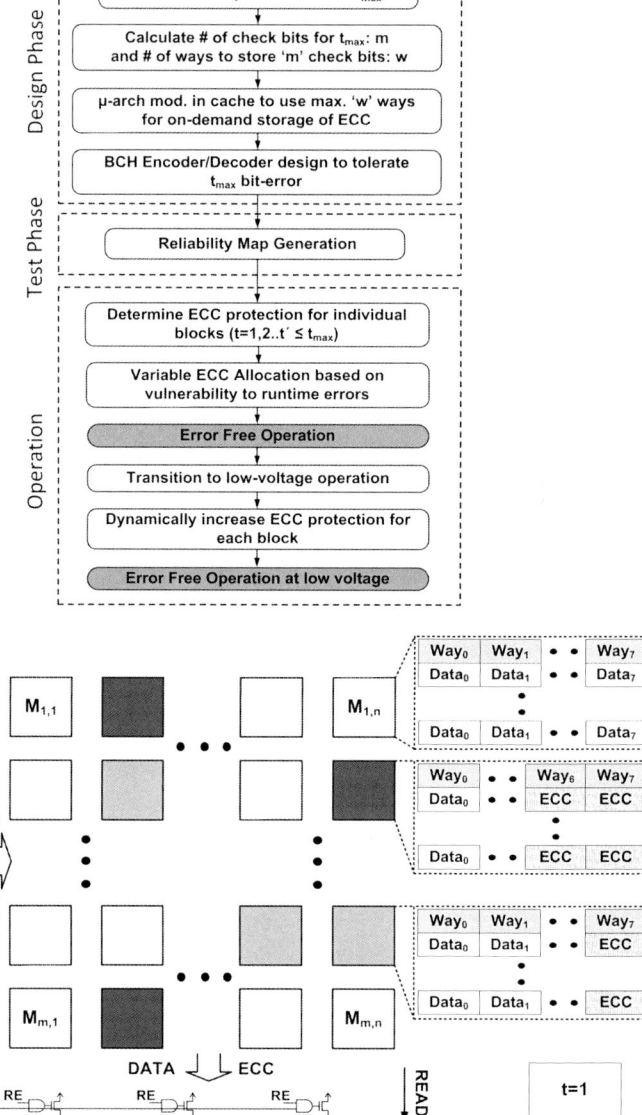

Fig. 19.2 (**a**) Major steps in variable ECC allocation; (**b**) Architecture for post-fabrication variable ECC allocation based on the process corner of the individual memory blocks

19.2 Reliability-Driven ECC Allocation for Multiple Bit Error Resilience

map is generated which characterizes the vulnerability of different sections of the cache to runtime errors. The baseline vulnerability is determined by the amount of parametric variation (measured in terms of read/write margins) present in the most reliable memory block. The baseline vulnerability is used to determine the value of t_{min} for the proposed ECC allocation scheme. For example in the simulations performed, for a bit failure probability of 10^{-6}, most reliable memory blocks have been assigned $t_{min} = 2$. Based on the reliability map, values of $t_{min} < t <= t_{max}$ for other memory blocks are determined by their relative vulnerabilities to parametric failures. For variable ECC allocation during deployment, these t values are encoded into 2-bits and are stored on-chip in a small non-volatile memory. Based on these encoded values, variable number of ways in each set are dedicated to storage of ECC bits by asserting appropriate select signals to the ECC selection logic. During runtime, the ECC selection logic (present for each memory block) is responsible for distinguishing the data and ECC bits for each set being read.

Figure 19.2b shows the corresponding L2 cache architecture to enable the proposed variable ECC allocation. For a given error rate, let the three levels of ECC protection assigned to a L2 cache be $t_{min} = 1$, $t_{med} = 2$ and $t_{max} = 3$. $t = t_{min} = 1$ denotes the scenario where a single error in a memory word (32-b or 64-b) can be safely corrected using conventional SECDED scheme. Typically the SECDED protection is applied to the L2 data arrays while runtime failures in the smaller tag arrays are typically avoided by either conservative design of the tag array or simple parity assignment [8]. Assuming 8 ECC bits per 64-b of L2 cache, SECDED scheme for 2MB L2 cache will require a storage of 256KB. Conventionally a separate storage is dedicated for storing these ECC bits on-chip [9]. In the variable ECC assignment scheme, multiple-bit error correction facility in less reliable memory blocks is achieved using BCH codes. For example, $t = 2$ denotes a 2-b error correction and 4-b error detection scheme. Similarly $t = 3$ denotes 3-b error correction and a 6-b error detection scheme, all achieved using BCH codes. Higher values of t require extra ECC bits. For example, for each 64-b word, the number of ECC bits required for $t = 2$ and $t = 3$ are 14 and 21, respectively. In order to accommodate these extra ECC bits, one or more of the cache ways are used to store ECC bits in place of actual data bits. This is illustrated in Fig. 19.2b. The derivation for the number of ways which needs to be devoted for storing the ECC bits is provided in [7].

The proposed scheme exploits the fact that modern L2/L3 caches typically have large number of "ways", some of which can be used to store check bits for on-demand error tolerance. Thus, contrary to the 2-D error coding scheme proposed in [10], allocating one or more ways in the same row for ECC storage does not cause unnecessary read from multiple rows. Figure 19.2 shows the encoder and decoder circuits for multiple-bit ECC protection. The encoder logic is used during the process of write (or store) to the L2 cache. At all other times, encoder logic corresponding to $t = 1, 2$ and 3 are gated off. Similar to an only SECDED scheme, any store shorter than the ECC word (64-b) requires a read-modify-write operation to the ECC stored in the way as well as to the ECC bits in the memory conventionally used for storing SECDED.

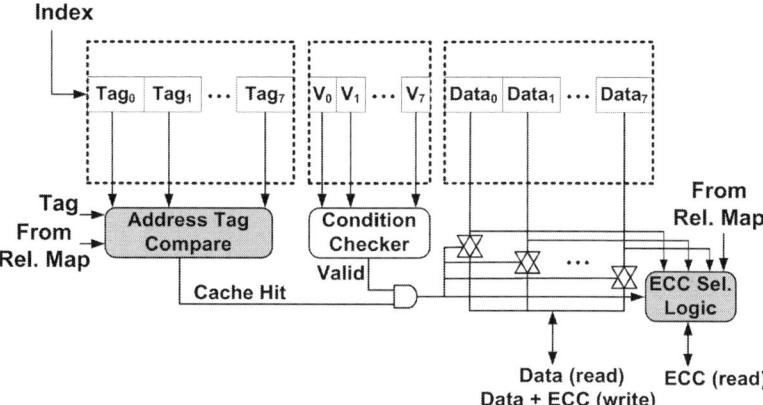

Fig. 19.3 μ-architecture showing procedure to store and retrieve ECC bits from one or more cache ways

Decoding for error correction involves two components: (i) logic for detection of error and (ii) finding out the error location in order to achieve error correction. For BCH encoding, the first component is denoted as *Syndrome detect*. The second component is further subdivided into *Berlekamp–Massey (BM) module* and *Chien Search module*. The correction circuit is invoked only if an error is detected. For realistic error rates, error-free read will be performed most of the time. Since the detection logic appears in the critical path for L2 read, a single-cycle decoding logic has been implemented for BCH. Moreover, the error correction logic is pipelined allowing a 64-b throughput per clock cycle. Details of this architecture which allows a post-fabrication ECC allocation is provided in [7].

19.2.2 μ-Architectural Modifications for Variable ECC Allocation

Figure 19.3 shows the modifications to the L2 cache μ-architecture required to realize the variable ECC assignment. In addition to the address tag, an enable signal from the reliability map would now serve as an input to the *Address Tag Compare* logic (Fig. 19.3). Given the memory blocks are classified into 4 or less ECC bins, this signal would be only 2-b wide. This actually serves to invalidate the tag comparison operation for one or more ECC ways so that during read, data out is not selected from any of these ways. If there is a hit in any of the remaining data ways, the cache hit signal and the enable signal coming from reliability map can be used to separate the ECC bits from the data bits using an *ECC selection logic* as shown in Fig. 19.3. During write, the ECC bits stored in these ways is modified based on the incoming data to be written. Write occurs to the cache way for which the tag hit

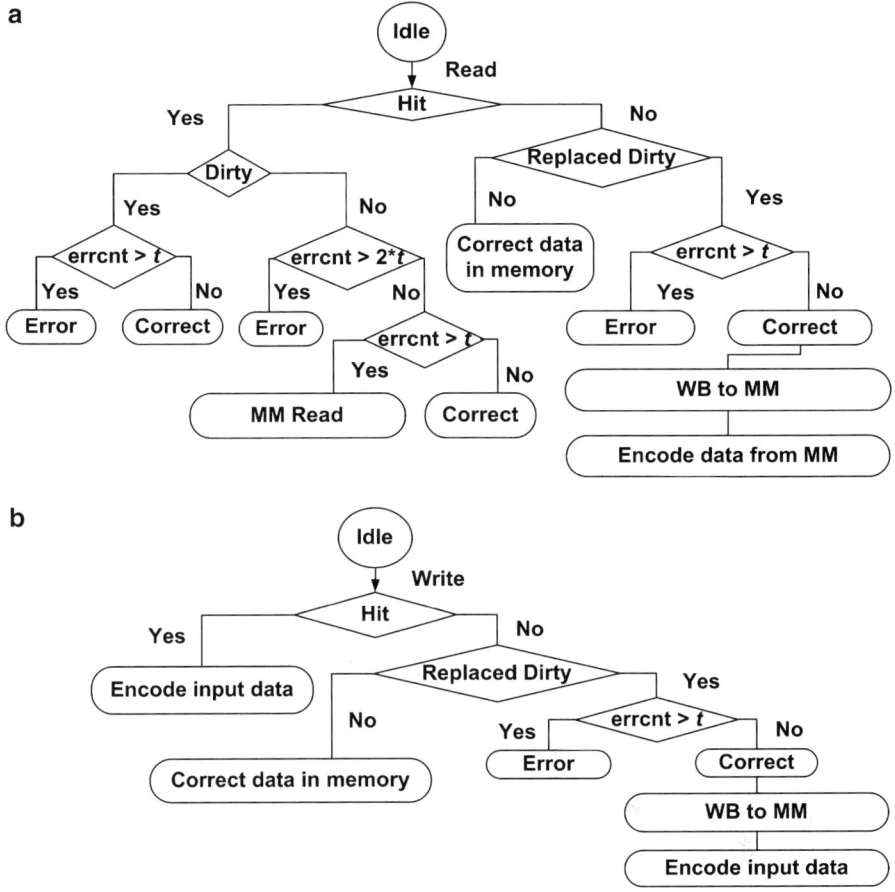

Fig. 19.4 States of the L2 ECC controller for error detection and correction

has occurred. In evaluating the performance for variable ECC, it may be noted the number of runtime errors in the L2 cache that are either transferred to the processor pipeline during loads or are transferred to the main memory in case of a read-miss or a write-miss. Figure 19.4 shows the L2 states during which the encoder, error detection and correction logic are invoked. Here a write-back L2 is assumed with a write-allocate scheme. Details of the controller states is provided in [7].

19.3 Experimental Results

The effectiveness of the proposed ECC allocation scheme was validated for the following scenarios: (1) tolerance to random failures; (2) tolerance to contiguous failures; (3) simultaneously tolerating both random and contiguous failures; and

(4) tolerating high failure rates at low voltage. Details of the simulation setup are provided in [7]. For evaluating the tolerance to runtime random errors, a weighted random pattern generation tool was integrated with the Simplescalar setup to insert random runtime errors into the L2 cache following a given bit failure probability (BFP). A BFP of 10^{-n} indicates a single bit failure out of 10^n bits. Considering a wide range of device parameter variations for PTM 45nm model, a large BFP range ($10^{-7} - 10^{-4}$) was observed in the SPICE simulations. It may be noted that a similar range has been observed in [11]. In the simulations, the Mean Time To Failure (MTTF) is taken to be 10^5 cycles. Hence, the error pattern is regenerated every 10^5 cycles. Note that the choice of MTTF has been made to facilitate the simulation process. A smaller or larger value of MTTF would not affect the average number of failures in a simulation window and hence is not expected to affect the choice of ECC or the energy/performance trends. The total number of errors in L2 which propagates into the processor pipeline or to the main memory was noted at the end of simulation. Energy values for the baseline processor and the proposed variable ECC scheme is obtained for 100nm CMOS process using Wattch [12].

19.3.1 Tolerance to Random Errors for BFP =10^{-4}

Although a wide range of *BFP* was simulated, this chapter presents the results for $BFP = 10^{-4}$. Simulation results for other values of *BFP* are provided in [7]. The baseline processor suffers from significant number of runtime failures for both worst as well as best-case reliability maps (refer to Fig. 19.5). For the proposed scheme, a $t = 2, 3$ and 4 distribution is able to correct *99.9%* and *99.8%* of all the failures that occur in the baseline processor. However, as shown in Fig. 19.5, it fails to account for all the errors. In order to ensure an error free operation, all the memory blocks must be protected with $t_{min} = 4$ for the die with the worst-case and $t_{min} = 3$ for the die with best-case reliability maps respectively. When the proposed scheme with $t_{min} = 4$ is used to protect all memory blocks in a die with worst reliability map, the average increase in CPI and energy are *3.58%* and *3.8%* respectively. For the best reliability map, the CPI and energy increase are *2.47%* and *2.53%* respectively (refer to Fig. 19.6). Average increase in L2 miss rate are *17.14%* and *17.12%* respectively.

19.3.2 Tolerance to Soft-Error Induced Contiguous Errors

For protection against soft error induced failures, it is assumed that the baseline processor is protected using SECDED along with 4-way bit interleaving. Such a system can correct up to 4 contiguous errors and detect up to 8. This suffices to cover the maximum number soft-error induced contiguous failures (4, 5 or 7) corresponding to $N_c = 16, 32$ or 64, where N_c refers to the number of memory cells between two well taps [3]. However, as pointed out in [10], bit interleaving comes at

19.3 Experimental Results

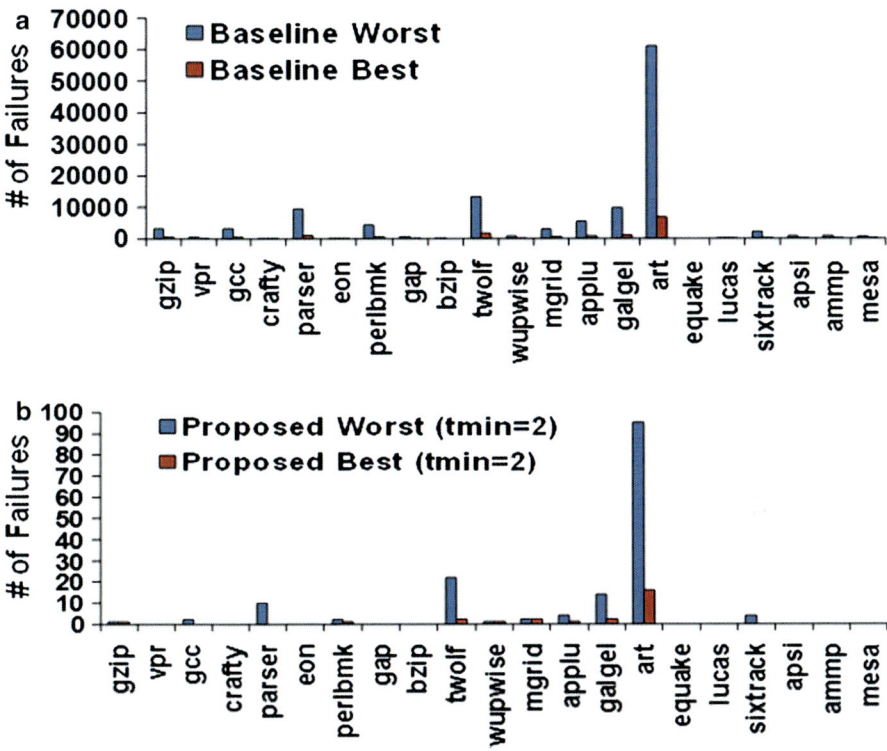

Fig. 19.5 Number of failures in (**a**) Baseline and (**b**) Proposed Scheme with BFP = 10^{-4}

additional power, delay and area overhead. Power overhead in an interleaved cache, which occurs primarily due to access of unwanted words in the same row has been shown to be more than *100%* of a non-interleaved one [10]. It is therefore interesting to see the trade-off involved in tolerating multiple-bit errors using on-chip BCH based ECC as opposed to bit-interleaving.

For memory designs with $N_c = 16$ or 32, an ECC distribution with $t = 4, 5$ and 6 suffices to cover the maximum number of contiguous errors that may occur. Penalty is however paid in terms of higher conflict miss and increased CPI overhead since 2–3 ways in each set are used for storing the ECC instead of data. Figure 19.7 shows the CPI overhead for the individual benchmarks for worst and best-case reliability maps. The average CPI overhead in these two cases are *3.52%* and *3.29%*, respectively. Although the L2 miss rate increases significantly, the variable ECC allocation actually leads to energy savings over the baseline processor which employs bit interleaving for tolerating multiple-bit contiguous errors. Total energy consumption for the baseline processor was estimated assuming the fact that bit interleaving doubles the energy requirement for each L2 access [10]. For the proposed scheme without bit interleaving, energy calculation considers the overhead due to $t = 4, 5$ and 6 ECC logic. Figure 19.7b shows the energy overhead for the

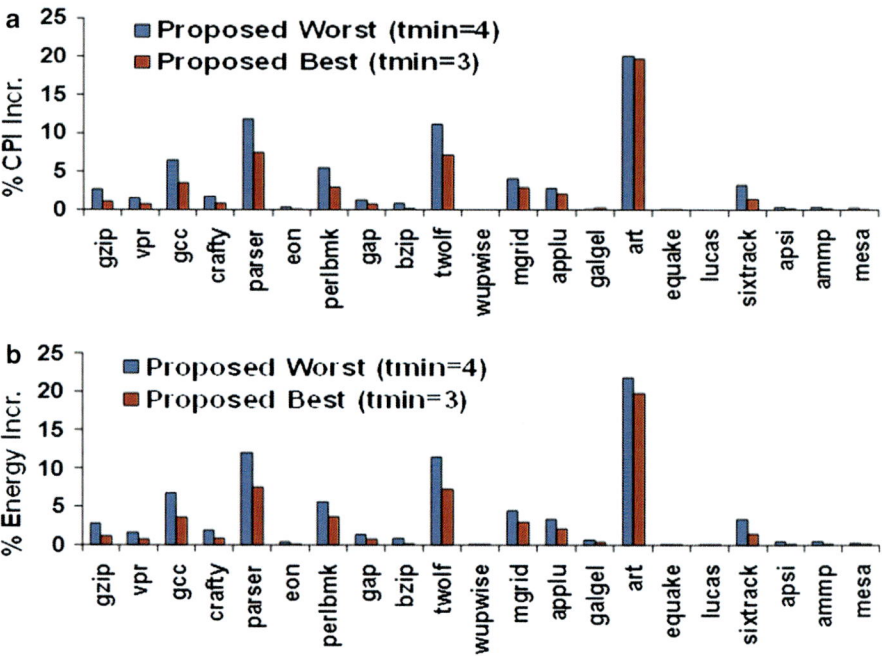

Fig. 19.6 Percentage increase in (**a**) CPI and (**b**) Energy for the proposed scheme with BFP $= 10^{-4}$

individual benchmarks. Average savings for the worst and best maps are 4.33% and 4.13%.

19.3.3 A Unified Solution to Tolerate Random and Contiguous Errors

Trade-offs exist with schemes that can serve as unified solutions to tolerating both random as well as contiguous errors in memory words. They are as follows.

19.3.3.1 $t_{min} = 4, 5$ and 6 Distribution with Extra Storage

Variable ECC allocation with $t = 4, 5$ and 6 distribution leads to inordinate increase in L2 miss rate for a set of benchmarks. This can be addressed if additional storage is available for storing the extra parity bits required by BCH encoding rather than using one or more ways to store the ECC bits. Assuming an additional storage for SECDED is already present, storage required for storing 20 extra parity bits corresponding to $t = 4$ is calculated to be 31% of the L2 area. For blocks with $t = 5$ and $t = 6$, it is assumed that the additional parity bits occupy 1 out of 8 ways in L2. This area increase is with respect to a L2 cache without bit interleaving

19.3 Experimental Results

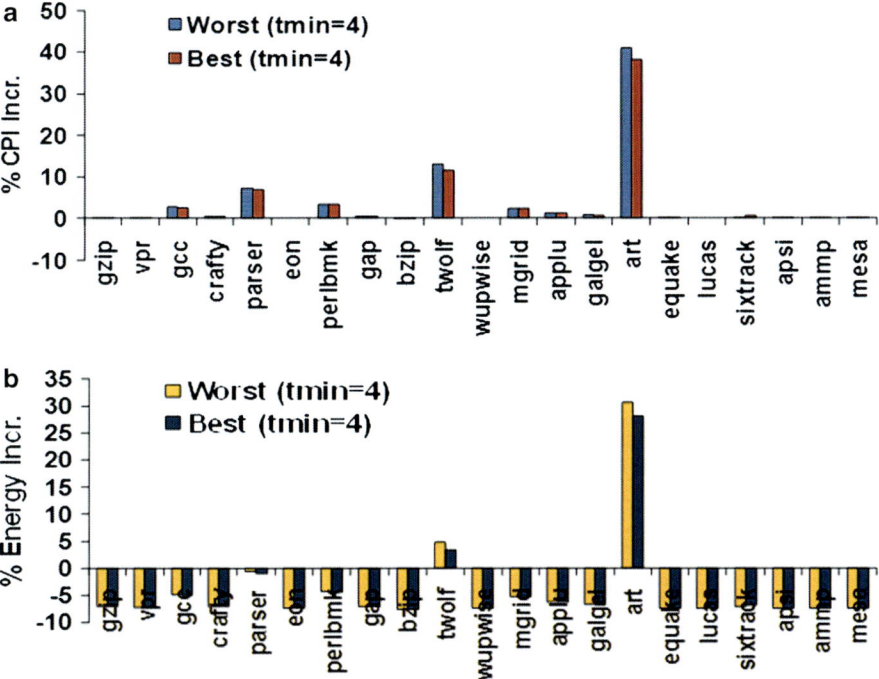

Fig. 19.7 Percentage increase in (**a**) CPI and (**b**) Energy when the proposed scheme is used for soft-error tolerance with $t_{min} = 4$

scheme. If this area overhead can be tolerated then, the average L2 miss rate increase is a minimal *3.68%*. After accounting for this additional energy overhead, a $t = 4, 5$ and 6 distribution with extra memory leads to an average energy savings of *6.64%*. Results for individual benchmarks are shown in Fig. 19.8b.

19.3.3.2 t_{min} = 2,3 and 4 Distribution with Bit-Interleaving

A variable ECC allocation with $t = 2, 3$ and 4 and 4 way bit interleaving tolerates both: (i) random errors with error rate as high as 10^{-4}; (ii) up to 8-bit contiguous errors in $t = 2$ blocks and up to 16-bit contiguous errors in $t = 4$ blocks, making it amenable to tolerance of high soft error rate and scalable to future technology nodes. With respect to a L2 cache without bit interleaving, this scheme does not incur any extra area overhead since extra ECC bits for $t = 2, 3$ and 4 are stored in ways. The average CPI overhead is only *1.56%* (refer to Fig. 19.8a). With respect to an only bit-interleaved L2, this scheme incurs an average *1.57%* increase in energy requirement (refer to Fig. 19.8). The average L2 miss rate however increases by *11.64%*.

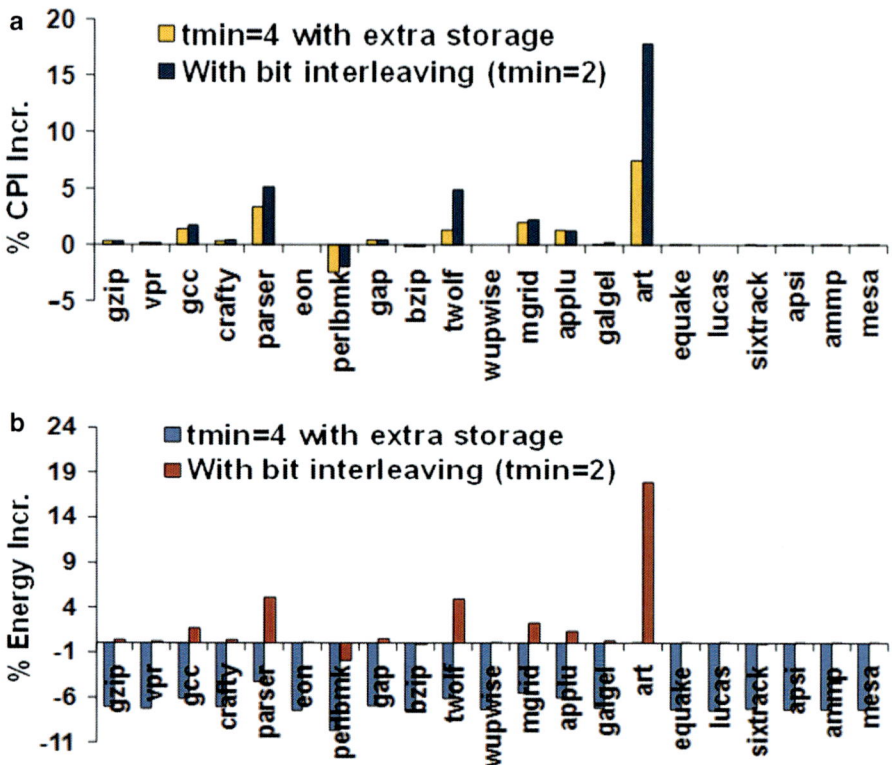

Fig. 19.8 Overhead in: (**a**) CPI and (**b**) energy when the proposed scheme with $t_{min} = 2$ is used along with bit-interleaving for soft-error tolerance

19.3.4 Energy Saving with Dynamic Voltage Scaling

The proposed multi-bit error tolerance scheme provides an opportunity to dynamically allocate ECC for the memory blocks. Such dynamic adaptation can be used to achieve power saving by combining voltage scaling for the memory cells with higher ECC protection. A 200mV reduction in supply voltage reduces the read, write and leakage powers by *55.86%, 33.15%* and *55.7%*, respectively.

This energy saving due to low power operation outweighs the overhead in providing extra ECC protection for $BFP = 10^{-4}$. Using Cacti the ratio of leakage to dynamic power is first estimated for a 2MB cache. For 100nm node this ratio is *1.03*. The total energy consumption including the leakage overhead for the L2 is then estimated for both nominal as well as low power operating conditions. After considering the energy overhead for $t = 4$ ECC protection for all memory blocks, average energy savings at 100nm node with the proposed scheme is calculated to be *5.93%* (Fig. 19.9). Higher savings in energy is achieved for 70nm technology node. At 70nm node, savings in read, write and leakage energy increases to *64.76%*,

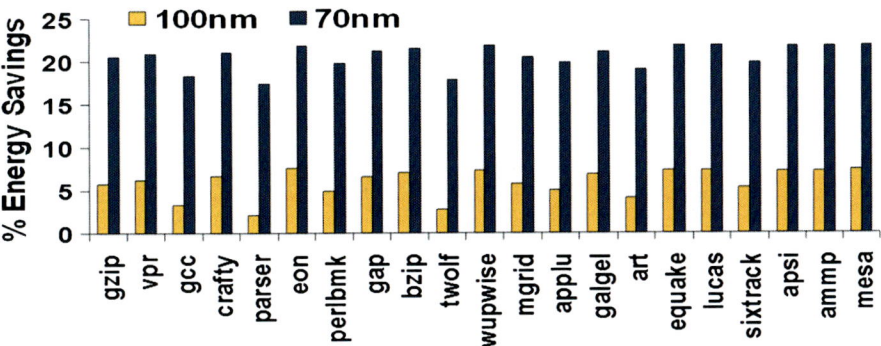

Fig. 19.9 Percentage energy savings at 100 and 70nm technology nodes when the on-chip L2 cache is operated in a low power mode with the variable ECC scheme accounting for the increased failure rate in low power mode

48.77% and 62.8%, respectively. The ratio of leakage to dynamic energy also increases to 5.17. Taking into account all these factors, the average energy savings at 70nm is 20.58%. Note that in order to provide increased protection during low power operation, the cache must go through a number of transitions before it fully operates in the low power mode. These actions ensure that correct ECC bits are stored for the memory blocks. Details of this transition is provided in [7].

References

1. J. Maiz, S. Hareland, K. Zhang, P. Armstrong, "Characterization of Multi-bit Soft Error events in advanced SRAMs", in *Intl. Electron Devices Meeting*, 2003
2. C.W. Slayman, "Cache and memory error detection, correction, and reduction techniques for terrestrial servers and workstations". IEEE Trans. Device Mater. Reliab. **5**(3), 397–404 (2005)
3. K. Osada, K. Yamaguchi, Y. Saitoh, "SRAM immunity to cosmic-ray-induced multierrors based on analysis of an induced parasitic bipolar effect". IEEE J. Solid State Circ. **39**(5), 827–833 (2004)
4. D.M. Kwai et al., "Detection of SRAM Cell Stability by Lowering Array Supply Voltage", in *Asian Test Symposium*, 2000
5. A. Pavlov et al., "Weak cell detection in deep-submicron SRAMs: A programmable detection technique". IEEE J. Solid State Circ. **41**(10), 2334–2343 (2006)
6. S. Mukhopadhyay, K. Kim, H. Mahmoodi, K. Roy, "Design of a process variation tolerant self-repairing SRAM for yield enhancement in nanoscaled CMOS". IEEE J. Solid State Circ. **42**(6), 1370–1382 (2007)
7. S. Paul, F. Cai, X. Zhang, S. Bhunia, "Reliability-driven ECC allocation for multiple bit error resilience in processor cache". IEEE Trans. Comput. **60**(1), 20–34 (2011)
8. S. Rusu, H. Muljono, B. Cherkauer, "Itanium 2 Processor 6M: Higher Frequency and Larger L3 Cache, in *Intl. Symposium on Microarchitecture*", 2004
9. J.L. Shin, B. Petrick, M. Singh, A.S. Leon, "Design and implementation of an embedded 512-KB Level-2 cache subsystem". IEEE J. Solid State Circ. **40**(9), 349–352 (2005)

10. J. Kim, N. Hardavellas, K. Mai, B. Falsafi, J.C. Hoe, "Multi-bit Error Tolerant Caches Using Two-Dimensional Error Coding", in *Intl. Symposium on Microarchitecture*, 2007
11. Z. Chisti, A.R. Alameldeen, C. Wilkerson, W. Wu, S. Lu, "Improving Cache Lifetime Reliability at Ultra-low voltages", in *Intl. Symposium on Microarchitecture*, 2009
12. D. Brooks, V. Tiwari, M. Martonosi, "Wattch: A Framework for Architectural-Level Power Analysis and Optimizations", in *Intl. Symposium on Computer Architecture*, 2000

Chapter 20
Summary

We have presented a novel hardware reconfigurable framework that utilizes two-dimensional memory array for reconfigurable computing. The idea proposed is to partition an application to multi-input multi-output LUTs and map them to dense memory arrays at each compute element of the framework. Each of these compute elements evaluates the LUTs over multiple clock cycles in a temporal manner and multiple compute elements combine to spatially map any large application.

We have defined the hardware architecture for the proposed spatio-temporal reconfigurable platform. Details of synchronization across multiple compute elements in the framework using time-multiplexed interconnect have been presented. The hardware modifications to realize a reconfigurable framework on demand from an existing memory organization are also proposed as part of this work.

In addition to the hardware architecture, we have presented a complete software flow which automates the application mapping process to the proposed MBC framework. The software flow not only allows design space exploration for the MBC framework, but also functional validation of applications mapped to this framework.

We have demonstrated how the memory based computing model can be effectively used to improve the reliability of operation for a general-purpose processor framework. In addition, we have also investigated hardware-software co-design approaches for improving the reliability of operation for the MBC framework.

Finally, we have presented MAHA, a memory based hardware reconfigurable acceleration framework, which instruments an off-chip CMOS-compatible non-volatile Flash memory for on-demand computing. Computing in such close proximity with the data minimizes the energy requirement for the system, , particularly for data-intensive applications, and hence, mitigates the Von-Neumann bottleneck.

We believe this work lays the foundation for alternative reconfigurable frameworks which are significantly more energy-efficient, scalable with technology and amenable to easy application mapping than existing ones. There are, however, a number of design choices and software optimizations which can be exploited to enhance the performance and functionality of the proposed MBC framework. We are listing some of them below:

- *Modifications to the MLB architecture:* The MLB architecture can be further optimized. Currently we use dummy registers to support feed-forward signals within each MLB. This is primarily because we do not support register selection while writing the result of a LUT or a datapath operation. If selection is allowed for both reading and writing to the temporary registers, then dummy registers can be eliminated.
- *Exploration of inter-MLB interconnect architecture:* A larger scope of optimization is present in defining the inter-MLB communication architecture. For MBC based stand-alone reconfigurable computing, we currently rely on an elaborate programmable interconnect framework between MLBs similar to the FPGA. We have demonstrated that MBC reduces the requirement of programmable interconnect significantly through localized execution and time-multiplexing of the inter-MLB communication channels. It is therefore possible to explore an inter-MLB interconnection framework, which is far less demanding in terms of area, delay and power compared to that in conventional FPGAs.
- *Software optimizations:* A major scope of optimization lies in the software tool set. The spatio-temporal nature of the proposed framework imparts unique advantages which cannot be leveraged in a purely spatial framework. For example, it is possible to minimize the impact of inter-MLB routing delay by scheduling the dependant operations in the next clock cycle. This retiming increases the latency by one clock cycle but prevents increase in cycle time. Similarly, to minimize the routing congestion, it is possible to schedule the transmission of signals over a time-multiplexed interconnect, which is again not possible over a fully spatial framework. All these optimizations can be part of future investigations.
- *Exploring the effectiveness of MAHA framework for on-chip memories:* We have noted that reconfigurable computing based on off-chip non-volatile memory brings significant improvement to system energy and relieves the bandwidth bottleneck for data-intensive applications. In future, one can investigate the effectiveness of on-chip MAHA architecture that uses embedded memory in a processor or system-on-chip in accelerating compute/data intensive applications.
- *Exploring the effectiveness of MBC framework for emerging non-volatile memories:* A number of promising non-volatile non-CMOS memory technologies are already on the horizon. For some of these technologies, we have already demonstrated that MBC is an extremely promising computing platform. Future work will explore other emerging memory technologies such as Memristor and PCRAM and investigate their benefits/challenges in realizing the proposed MBC framework.

Printed by Publishers' Graphics LLC